CONTENTS

FORESTRY COMMISSION

Booklet 40

Chemical Control of Weeds in the Forest

by
R. M. Brown, B.Sc.
Forestry Commission

LONDON: HER MAJESTY'S STATIONERY OFFICE

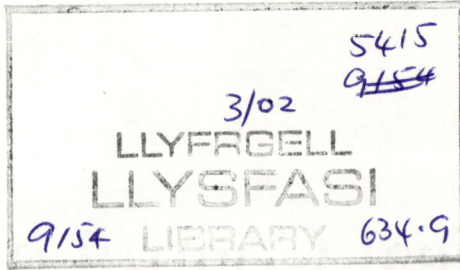

ISBN 0 11 710185 0

CHAPTER 1

GENERAL INFORMATION AND SUMMARY RECOMMENDATIONS

INTRODUCTION

1.1.   This booklet gives recommendations on the use of herbicides in British forests. It replaces Forestry Commission Leaflet 51 (Aldhous, 1969) also entitled *Chemical Control of Weeds in the Forest*.

1.2.   Suitable types of equipment are generally described in Chapter 8, but detailed advice on the most suitable type and make of spraying equipment is not given as this may be found in Forestry Commission Bulletin 48, *Weeding in the Forest* (HMSO £2·10) (Wittering, 1974). Recommendations for the use of herbicides *in forest nurseries* may be found in Forestry Commission Bulletin 43 entitled *Nursery Practice* (HMSO £1·50) (Aldhous, 1972), or in "*Volume II, 7th Edition, Weed Control Handbook*" (Fryer and Makepeace, 1972).

1.3.   In the sections giving detailed advice on the doses and times of application of herbicides for particular weed problems, a distinction is made between firm recommendations, provisional recommendations and purely information sections on new herbicides. The essence of each firm recommendation is given in bold type at the beginning of each paragraph concerned with a particular use of a herbicide. Any qualifications or information relating to this recommendation follow in normal print in sub-paragraphs. Provisional recommendations for uses of a herbicide which are not well proven commence with "(Provisional)", and are in plain type. Similarly, paragraphs giving information on a new herbicide commence with "(Information)" and are in plain type.

1.4.   All recommendations for doses are made in terms of weight of product per hectare treated, but the percentage content of the active ingredient or acid equivalent (see para 1.16) in the product is given, and the weight of active ingredient or acid equivalent per hectare given in brackets. For herbicides applied as sprays, the amount of diluent required per hectare sprayed is given using the general terms medium, low or ultra low volume (see Appendix B. Glossary of Technical Terms).

1.5.   The experimentation and experience which forms the basis of these recommendations has been mainly in plantations of the coniferous species commonly used in Britain, or in broadleaved woodland being converted to such plantations. A very large proportion of the species planted every year in Britain have been conifers, but because broadleaved trees are becoming increasingly important, in this booklet an attempt has been made to draw attention to recommendations which are suitable for broadleaved plantations.

1.6.   Unless it is otherwise stated, all recommendations for pre-planting applications are equally applicable to areas to be planted with conifer or broadleaved species. It has been necessary to be more specific for post planting applications, and the recommendations for each herbicide make it clear what limitations there are on the crop species that can be treated.

WHEN TO USE HERBICIDES

Technical Considerations

1.7.   The use of herbicides can be justified as a means of saving money on establishing a crop, or of reducing seasonal peak labour requirements. The materials mentioned below can be recommended for one or both these reasons. Sometimes their use will also lead to more rapid growth of the young crop, but this cannot be put forward as a benefit to be obtained in the majority of occasions when a herbicide is used for weed control. Sometimes the use of a herbicide will be prevented for environmental reasons, in spite of the fact that its use may give economic or managerial advantages.

1.8.   No herbicide can be relied on always to give complete kill of the weeds. This may be because either the dose has to be small enough to avoid crop damage, or to the presence of weeds which are resistant to the herbicide. Often patches of weed survive, or one or two sprayed stumps sprout again because of the inherent difficulty of spraying evenly where weed growth is patchy and irregular, and in other cases to the protection of small weeds by bigger neighbours so that the herbicide does not reach its target properly. Such missed areas may require "spot" or local re-treatment. On the other hand a complete kill of weeds is rarely necessary to prevent undesirable competitive effects on the crop.

1.9.   Weed flora in forest situations are usually a mixture of many species. Rarely does any chosen herbicide control all these, and there are situations in which weeds resistant to any herbicide suitable for use in forest crops are so frequent that hand or mechanical methods of weed control must be used.

1.10.   New vegetation may replace the killed weed flora; for example, grasses may increase rapidly when broadleaved weeds such as bramble are controlled. Before spraying, the likely consequences of removing any particular weed or group of weeds should be carefully considered.

1.11.   For the best results the use of herbicides must be fully integrated into the process of establishing a crop or removing an unwanted vegetation cover, and not treated as an isolated operation. Herbicides should never be regarded as the *exclusive* way of controlling weeds. Machine cutting, especially with tractor-powered machines, is often cheaper and excellent results can be obtained by a combination of machine and chemical control, or of hand cutting and chemical control.

Safety Considerations

1.12.   It would be a disservice to suggest that there are no hazards associated with the use of herbicides in forestry. On the other hand, herbicides have been used on a practical scale in British forests for the last 15 years, and during this period no significant illness properly attributable to any use of a herbicide as recommended in this booklet has been recorded. This good safety record is partly due to the care taken during the handling of herbicides, and it is important that the same care is always taken in the future. Herbicides should be considered as tools which can be properly used or abused. If the instructions for use are not strictly followed, extensive damage to plants or discomfort to users may follow.

1.13.   All the herbicides fully recommended in this booklet have been cleared under the Pesticides Safety Precautions Scheme (see para 11.3–11.12). They should, therefore, be marketed in containers on which the labels have been carefully drawn up to give all the essential information on the precautions to be taken and the protective clothing to be worn when using the contents.
   It is essential to keep herbicide concentrates in their containers and to read, understand and act on all the information on the label.

1.14.   In the section in this booklet on safety precautions (see paras 11.22–11.31) the rather generalised recommendations made under the Pesticides Safety Precautions Scheme are interpreted into practical recommendations on the precautions to be taken when using herbicides in forest situations.
   All supervisors should be aware of the hazards and the precautions to be taken when using a herbicide, and that they are responsible for ensuring that the operator is taking the correct precautions.

2

CHOOSING A HERBICIDE

Table Summarising the Recommendations in this Booklet

1.15.   Table 1 gives, for each weed type, the herbicides which are fully recommended, the main methods of application and the times of year when the method is recommended for use. The recommended months of application are a compromise, combining degree of kill, susceptibility of crop species and ease or convenience of application. Detailed recommendations are to be found in the text by referring to the "Contents" pages at the front of the booklet.

Choice of Formulation

1.16.   Compounds showing herbicidal activity are rarely suitable for use in their simplest form (ammonium sulphamate is an exception). The process of combining a herbicidal compound with other compounds or mixing it with surfactants, emulsifiers, stickers etc. to make it easier to apply and more effective in adhering to or penetrating plant tissue is called *formulation*. The herbicidal compound in any formulation is called the *active ingredient* (a.i.) or, if it is active as an acid, *acid equivalent* (a.e.). In everyday parlance, the commercially sold herbicide is the herbicide formulation, and the container label should always show how much active ingredient or acid equivalent it contains.

1.17.   The formulation may affect many important properties of the herbicide, in particular its ability both to contact and to penetrate the outer tissue of the plant, its ability to mix with the recommended diluent, its volatility and its cost. Where more than one type of formulation is available, the recommendations for the use of a herbicide in this leaflet make it clear which formulation is to be used, and give reasons for the choice.

Naming of Recommended Herbicides

1.18.   In the following paragraphs, the common name of each herbicide is used. The correct common name can be found in British Standard 1831 (latest edition 1969) or Volume I, 5th Edition (revised) of the *Weed Control Handbook*, (Fryer and Evans, 1970). Details of the proprietary brand names and the suppliers of most of the herbicides mentioned can be found in the current *Agricultural Chemicals Approval Scheme—List of Approved Products*. This is published annually in February by the Ministry of Agriculture, Fisheries and Food and is available free of charge from the Ministry's Publications Department, Block C, Tolcarne Drive, Pinner, Middlesex.

3

SUMMARY OF RECOMMENDATIONS FOR
PRE-PLANTING TREATMENT

WEED TYPE	HERBI-CIDE	APPLI-CATION TYPE	TIME OF APPLICATION	DOSE PRODUCT/HECTARE (EQUIPMENT)
Grasses and grass/herb mixtures	atrazine	foliar	(MAY ... APRIL)	8–12 kg 50% w/w powder in water at MV (K, PH) or 100–150 kg 4% w/w granules (GD) (not more than 8 kg powder or 100 kg granules for N.S., W.H. and E.L.)
	dalapon	foliar	not immediately before or after rain; not within 3 weeks of planting	17 kg 74% w/w powder in water at MV (K, PH)
	chlor-thiamid and dichlo-benil	surface	leave one month before planting	40–60 kg 7½% w/w granules (GD)
	paraquat	foliar	molinia only before 1st Sept; not within 3 days of planting	5–10 litres 20% w/v concentrate in water at MV (K, PH)
Woody weeds and mixtures of woody and herbaceous weeds	2, 4, 5–T	foliar	gorse and broom can be killed in winter	7 litres 50% w/v concentrate in water at LV (AS, M) or in water at MV (K, PH) or 10 litres 30% w/v special ULV formulation (ULV)
	2, 4, 5–T/ 2, 4–D mixture	foliar		4.5 litres 75% w/v mixed concentrate in water at LV (AS, M) or in water at MV (K, PH)
	2, 4, 5–T	cut stump basal bark and frill girdle	spray only when dry—bark to run-off must be saturated; spray a month before planting if possible	2–3 litres 78% w/v concentrate per 100 litres paraffin (K, PH)
	2, 4–D Amine or 2, 4, 5–T (undiluted)	tree injection		1 ml 50% w/v 2, 4–D or 2, 4, 5–T concentrate at 75 mm centres (50 mm for resistant species) (T.I.)
	A.M.S. (ammonium sulphamate)	standing stem—frills		0.4 kg A.M.S. per litre water (plastic watering can) 15 g dry crystals per notch (notches 10 cm apart, edge to edge)
	A.M.S. (ammonium sulphamate)	cut stump	apply to freshly cut stumps within 24 hours of cutting; not within 12 weeks of planting	0.4 kg A.M.S. per litre water (plastic watering can or bucket-and — paint-brush) or 6 g dry crystals/cm of stem diameter
Heather	2, 4–D ester	overall		K, PH—10 litres 50% w/v concentrate in water at MV (K, PH) or in water at LV (AS, M) or 15 litres 40% w/v special ULV formulation (ULV)
Bracken	asulam	foliar	spray 24 hours before expected rain	10 litres 40% w/v concentrate in water at LV (AS, M)
	dicamba	foliar or surface	leave 3–4 months in summer and 4–6 months in winter before planting	M. 8–10 litres 40% w/v concentrate in water at LV (M) K. or 8–10 litres 40% w/v concentrate in water at MV (K, PH)
Rhododendron	A.M.S. (ammonium sulphamate)	cut stumps and regrowth	apply to freshly cut stumps to saturation point within 24 hours of cutting; leave for 12 weeks before planting	0.4 kg A.M.S. per litre water (plastic watering can or bucket-and-brush)
	2, 4, 5–T	cut stumps and regrowth	spray between March and October only if no sensitive crops in area	2.5–3.2 litres 78% w/v concentrate in 100 litres paraffin or 4.5 litres 50% w/v concentrate in 100 litres water or paraffin (K, PH)

FULL WIDTH OF SHADING INDICATES TIME OF APPLICATION GIVING A COMBINATION OF BEST KILL AND GREATEST SUITABILITY

DIMINISHING WIDTH OF SHADING INDICATES PROGRESSIVELY LESS SUITABLE OR LESS EFFECTIVE TIME OF APPLICATION

CUT-OFF INDICATES TREATMENT HAS TO TERMINATE, BECAUSE OF PLANTING OR ONSET OF DORMANCY

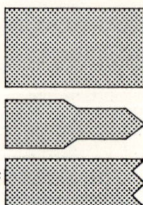

KEY: AS. AERIAL SPRAY (ONLY USED ON 'SPECIAL' OCCASIONS BY THE FORESTRY COMMISSION)
GD GRANULE DISTRIBUTOR
K KNAPSACK SPRAYER
PH LIVE REEL SPRAYER
TI TREE INJECTOR
M MISTBLOWER (KNAPSACK OR TRACTOR)
ULV ULTRA LOW VOLUME SPRAYER
LV LOW VOLUME
MV MEDIUM VOLUME

LE 1.

THE USE OF HERBICIDES IN FORESTRY

POST-PLANTING TREATMENT

WEED TYPE	HERBICIDE	APPLICATION TYPE	TIME OF APPLICATION	DOSE PRODUCT/HECTARE (EQUIPMENT)
Grasses and grass/herb mixtures	atrazine	foliar		8–12 kg 50% w/w powder in water at MV (K, PH) or 100–150 kg 4% w/w granules (GD) (not more than 8 kg powder or 100 kg granules for N.S., W.H. and E.L.)
	dalapon	foliar	spray only while crop is dormant / direct spray to miss crop foliage	13.5 kg 74% w/w powder in water at MV (K, PH + guard)
	chlor-thiamid and dichlobenil	surface	on L.P., C.P., S.P., S.S., N.S. oak, ash, beech and sycamore only	40–60 kg 7½% w/w granules (GD)
	paraquat	foliar	molinia only at this time	5 litres 20% w/v concentrate in water at MV (K, PH + guard)
Woody weeds and mixtures of woody and herbaceous weeds	2, 4, 5–T	foliar	apply after conifers have ceased height growth, but not in hardwoods, larches, Pinus contorta P.pinaster or P.radiata unless with placed knapsack application	4.5–7 litres 50% w/v concentrate in water at LV (AS, M) or in water at MV (K, PH) or 7 litres 30% w/v special ULV formulation (ULV)
	2, 4, 5–T/2, 4–D mixture	foliar		3–4.5 litres 75% w/v mixed concentrate in water at LV (AS, M) or in water at MV (K, PH)
	2, 4, 5–T	cut stump basal bark and frill girdle	late spring/summer treatment requires care	2–3 litres 78% w/v concentrate in 100 litres paraffin (K, PH)
	2, 4–D or 2, 4, 5–T (undiluted)	tree injection		1 ml 50% w/v 2, 4–D or 2, 4, 5–T concentrate at 75 mm centres (50 mm for resistant species) (TI)
	A.M.S. (ammonium sulphamate)	standing stems-frills		0.4 kg A.M.S. per litre water (plastic watering can) or 15 g dry crystals per notch (notches 10 cm apart, edge to edge)
	A.M.S. (ammonium sulphamate)	cut stump	occasional trees may be damaged	0.4 kg A.M.S. per litre water (plastic watering can or bucket-and-paint-brush) or 6 g dry crystals/cm of stem diameter
Heather	2, 4–D ester	overall	see footnote / not in hardwoods	8 litres 50% w/v concentrate in water at MV (K, PH) or in water at LV (AS, M) or 10 litres 40% w/v special ULV formulation (ULV)
Bracken	asulam	foliar	at least 24 hours should elapse before rain use only with S.P., C.P., S.S., N.S., D.F., G.F., J.L., Be and Bi	7–10 litres 40% w/v concentrate in water at LV (AS, M)
	dicamba		not suitable for post-planting use	
Rhododendron	A.M.S. (ammonium sulphamate)	cut stumps and regrowth	not suitable for post-planting use	
	2, 4, 5–T	cut stumps and regrowth	spray only in crop's dormant season	4–5 litres 50% w/v concentrate in 100 litres water (or paraffin) (K, PH)

Time of application column headings: MAY JUNE JULY AUG SEPT OCT NOV DEC JAN FEB MAR APRIL

FOOTNOTE

		TIMES OF APPLICATION OF 2, 4–D IN HEATHER	
Equipment	Species	North Britain*	South Britain*
M, ULV and AS	Spruces and Pines	mid July/early Sept	early Aug/mid Sept
	Other species	early Aug/early Sept	mid Aug/mid Sept
K, PH	All species	mid May to mid Sept (Crops under 1 metre mean height not to be sprayed in growing season)	

*North Britain lies north of the Mersey/Humber estuaries.

For key to equipment abbreviations see footnote under right-hand column of facing page

RECOMMENDATIONS I: PERENNIAL GRASSES AND GRASSES MIXED WITH HERBACEOUS BROADLEAVED WEEDS

GENERAL CONSIDERATIONS

2.1.   Perennial grass weeds, or mixtures of grass and herbaceous broadleaved weeds, frequently compete with newly planted tree crops during the first five years after planting. Only occasionally has the competition for moisture and nutrients by these weeds been shown seriously to retard crop growth, the main risk to the crop arising from excessive shading and "smothering" from weeds that are taller than the crop and often fall over it at the end of the growing season as the aerial parts die away. "Smothering" appears to damage crops mechanically and by creating conditions favourable for pathogenic fungi, as well as by creating excessive shade.

2.2.   As improvements in crop growth due to removal of weed competition for moisture and nutrients are unpredictable and often small, the main object of weed control in these situations is to prevent smothering. Unfortunately, the variation in weed floras and their vigour from site to site, and the variation in tolerance of smothering between crop species, makes it impossible to quantify the extent of weed control required to avoid smothering. Foresters must use their own experience of their area to assess the degree of weed control required.

2.3.   Five herbicides, atrazine, chlorthiamid, dalapon, dichlobenil and paraquat, are fully recommended for grass and grass/herbaceous broadleaved weed mixtures. Each one differs in the spectrum of weeds it controls well. The choice of herbicide depends initially on its ability to leave the crop undamaged, but thereafter the choice depends largely on the differences in weed species controlled and the cost. If herbaceous broadleaved weeds are really a problem, then herbicides which mainly control grasses (dalapon and atrazine) should not be used. However, in most situations, grasses form the most important and potentially dangerous fraction of the weed flora, and the particular species of grass present will largely dictate which herbicide should be used. Table 2 shows the susceptibility of the major grass species found in British forestry to these five fully recommended herbicides. It should be noted that the performance of each herbicide will vary with soil type, weather conditions and time of application.

2.4.   Figure 1 is a decision tree which should help foresters decide which herbicide to use for their particular weed situations. The final decision may often be a compromise, and will often depend on the major weeding type in a forest.

CONTROL OF GRASS AND GRASS/HERBACEOUS BROADLEAVED WEEDS BEFORE PLANTING

2.5.   Control of grass and herbaceous broadleaved weeds before planting is not commonly practised in Britain. New ground is often ploughed before planting and this gives adequate initial suppression of the weeds, whilst on ground to be replanted, the previous crop has frequently prevented the development of weeds. Even when weeds are present, these can be more conveniently controlled after planting rather than before.

2.6.   However, in some situations the control of existing weeds before planting can make the planting and establishment of the crop easier. In these situations it is rarely necessary to apply the herbicide to the

TABLE 2

SUSCEPTIBILITY OF COMMON GRASSES TO RECOMMENDED HERBICIDES

Grass Species Botanical Name (Common name)	Paraquat 1.1 kg a.i./ha	Dalapon 10·0 kg a.i./ha	Chlorthiamid/ Dichlobenil 4·5 kg a.i./ha	Atrazine 4·5 kg a.i./ha	General grass type
Agropyron repens (Couch)	MR	MS	MS	MR	soft
Agrostis gigantea (Black bent)	MS	MS	MS	MS	fine
Agrostis species	MS	S	S	S	fine
Anthoxanthum odoratum (Sweet vernal-grass)	MS	S	MR	S	soft
Arrhenatherum elatius (False oat-grass)	MS	S	S	MR	coarse
Calamagrostis epigejos (Wood small reed)	R	MR	MR	R	coarse
Dactylis glomerata (Cocksfoot)	MR	MS	MS	R	coarse
Deschampsia caespitosa (Tufted hair-grass)	MS	MS	MS	R	coarse
Deschampsia flexuosa (Wavy hair-grass)	S	S	S	S	fine
Festuca arundinacea (Tall fescue)	MS	MS	S	MS	fine
Festuca pratensis (Meadow fescue)	MS	MS	S	MS	fine
Festuca ovina (Sheeps fescue)	MS	MS	S	S	fine
Festuca rubra (Red fescue)	MS	MS	S	S	fine
Holcus lanatus (Yorkshire fog)	MS	S	S	S	soft
Holcus mollis (Creeping soft-grass)	MR	MS	MS	R	soft
Molinia caerulea (Purple moor-grass)	S	S	MS	R	coarse
Poa annua (Annual meadow-grass)	MS	MS	S	S	soft
Poa pratensis (Meadow-grass)	MS	MS	S	MS	soft
Poa trivialis (Rough meadow-grass)	MS	MS	S	S	soft

Notes: S = Susceptible: control should be excellent
MS = Moderately Susceptible: control should be adequate
MR = Moderately Resistant: control often inadequate
R = Resistant: little effect or control obtained

on the basis of satisfactory weed control persisting for one growing season after application of the herbicide.

CHOOSING A HERBICIDE TO CONTROL GRASS OR GRASS/HERBACEOUS
BROADLEAVED WEEDS ON PLANTED AREAS OR AREAS TO BE PLANTED
SHORTLY AFTER SPRAYING

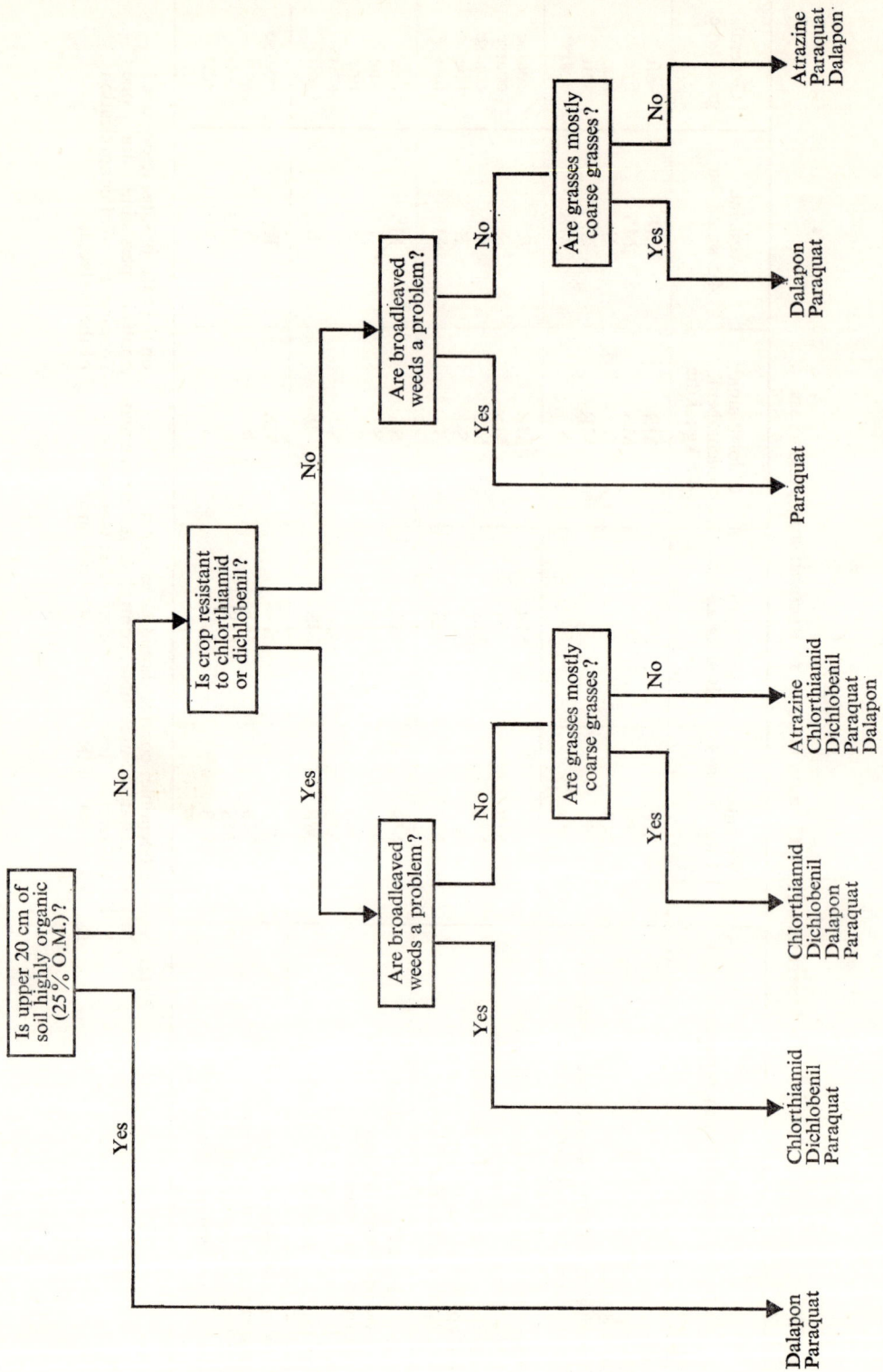

Note: Where more than one herbicide is recommended, they are listed in order of decreasing effectiveness

Figure 1. Algorithm to help choose the correct herbicide for controlling various grass and herbaceous broadleaved weed floras.

total area. It is usually better to apply the herbicide to metre wide strips or metre square patches along the future planting lines, and so reduce the cost of the herbicide per gross hectare. Applications should be made sufficiently in advance of planting so that the weeds in the sprayed strips or patches have begun to die and planting positions can be located.

2.7. DALAPON as 17 kg of 74% w/w commercial product per hectare (12·5 kg a.e./ha) in water at medium volume to grass weeds whilst they are growing reasonably vigorously, and no nearer than 3 weeks before planting so as to avoid crop damage.

2.7.1.   Best results are usually obtained if application is made about six weeks before planting as this gives time for the weeds to die down and make planting easier. Applications should be made in September or October for autumn planting, or in March or April for spring planting. Applications should not be made immediately before or after rain as dalapon is easily washed off the foliage.

2.7.2.   Dalapon has little effect on broadleaved weeds, and should only be used where grasses are the main problem. It is probably the most effective of the fully recommended herbicides for rhizomatous grasses and species of *Calamagrostis*. For deep rough swards a more complete kill is obtained if the sward is burnt, grazed or cut to encourage fresh growth before spraying. Control may also be poorer if new growth is shielded from the direct spray by dead material. Dalapon is sufficiently well absorbed through grass foliage to be active on any soil type, even peats.

2.7.3.   Dalapon formulations are usually available with or without a wetting agent. Formulations with a wetting agent are usually slightly more effective than those without.

2.8. PARAQUAT as 5 litres of 20% w/v concentrate per hectare (1·0 kg a.i./ha) in water at medium volume to green grass and herbaceous broadleaved weeds up to 2 or 3 days before planting.

2.8.1.   Applications are best made sufficiently in advance of planting to allow the weeds time to die down, and so make planting easier. In cool, dull autumn or spring weather, paraquat applications may take up to two weeks to show their full effect, although often the kill of weeds is more complete than when applications are made in hotter, brighter weather. Even so, paraquat is rarely translocated well enough within the plant to prevent many perennial weeds (especially grasses) shooting again in the first growing season after planting, and applications should not, therefore, be made too far in advance of planting.

2.8.2.   Paraquat must be applied to green tissue. Dead grass or other weeds will absorb the herbicide and may effectively screen green leaves. Since paraquat is not well translocated through plant tissues, good coverage of the weeds is essential. For vigorous, deep swards the dose of paraquat should be increased to 10 litres of 20% w/v concentrate per hectacre (2·0 kg a.i./ha), and a greater volume of water used (suggest 400 to 700 litre/ha). Since paraquat depends entirely on foliar contact and absorption for its activity, it is effective on all soil types, including peats.

2.8.3.   Always use low pressure, medium volume applications directed down onto low lying weeds so as to avoid any risk of drift of fine droplets. Paraquat must *never* be applied through a mistblower or ultra-low-volume sprayer as it is considered impracticable to protect the operator's eyes and respiratory passages from the fine droplets produced by such applications.

2.9.   (Provisional): Atrazine

Atrazine applied as 8 to 12 kg 50% w/w powder in water at medium volume, or as 100 to 150 kg 4% w/w granules per hectare (4–6 kg a.i./ha), may also be suitable for pre-planting weed control in situations where fine and soft grasses are the main problem. This herbicide has not been adequately tested, however, for this purpose, although its use *post-planting* is well proven (see para 2.17).

2.10. (Provisional): Chlorthiamid

Chlorthiamid applied as 40 to 60 kg $7\frac{1}{2}\%$ w/w granules per hectare (3·0 to 4·5 kg a.i./ha) applied up to one month before planting has been successfully used as a pre-planting treatment where the crop is to be of a tolerant species (see para 2.18.3). Applications earlier than February may not provide control for all of the subsequent growing season.

2.10.1.   Knowledge and experience of pre-planting applications of chlorthiamid is limited. However, crop and weed sensitivities and optimum application times are expected to be the same as for post-planting applications (see para 2.18).

2.11. (Provisional): Dichlobenil

Dichlobenil applied as 40 to 60 kg $7\frac{1}{2}\%$ w/w granules per hectare (3·0 to 4·5 kg a.i./ha) applied up to one month before planting is expected to give similar results to an identical chlorthiamid treatment (see para 2.10).

2.11.1.   Trials and experience of pre-planting applications of dichlobenil are even more limited than with chlorthiamid. The proposed crop should be of a species resistant to chlorthiamid (see para 2.18.3).

2.12. (Information). Glyphosate

Glyphosate applied as 7 to 13 litres of 36% w/v concentrate per hectare (2·5 to 5·0 kg a.e./ha) applied in March or June in medium volume has also shown considerable promise for controlling grass and herbaceous broadleaved weeds and may be suitable for pre-planting weed control treatments (see para 2.22).

2.13. (Information). Propyzamide

Propyzamide applied as 4 to 8 kg 50% wettable powder per hectare (2 to 4 kg a.i./ha) applied in November or December in water at medium volume has recently shown promise for controlling mainly grass weeds prior to planting.

2.13.1.   The optimum time of applying propyzamide to control perennial weeds in forest situations appears to fall between October and January in Britain. In 1972/73 trials conifers planted in the spring following applications in November and January were not damaged.

CONTROL OF GRASSES AND GRASS/HERBACEOUS BROADLEAVED WEEDS AFTER PLANTING Front cover and Plates 1–2.

2.14.   Grasses and herbaceous broadleaved weeds are more commonly controlled after planting than before. Even where weeds have been controlled before planting (by herbicides or ploughing) or where the previous crop suppressed most of the weeds, invasion of the site can be so rapid that some weeds may have to be controlled in the first season after planting.

2.15.   In many plantations on weedy sites, especially in lowland areas of south and east Britain, weeds may have to be controlled for three to five years or more after planting. The application of the appropriate herbicide on susceptible weed floras will often provide cheaper and more effective control than the use of hand cutting tools. However, weeding with cutting machines, especially tractor-powered machines, can be cheaper than weeding with herbicides in certain situations. For comparisons between the costs of using various herbicides see Chapter 10. For comparisons between the costs of using herbicides and other methods see Forestry Commission Bulletin 48: *Weeding in the Forest* (Wittering,1974).

2.16.   Trees are currently spaced at 2·1 × 2·1 metres in most plantations, and it is wasteful and usually unnecessary to treat the total area. Normally 1·0 to 1·3 metre wide strips along each row or 1·0 to 1·3 metre diameter patches round each tree are sprayed, depending on the vigour and expected height growth of the weeds. Inter-row application may be made in particularly vigorous weed situations.

2.17. ATRAZINE as 8 to 12 kg 50% w/w wettable powder in water at medium volume, or as 100 to 150 kg 4% w/w grannules per hectare (4 to 6 kg a.i./ha), applied from February to May in most conifer and many broadleaved tree crops.

2.17.1.   Atrazine is very effective against most fine and soft grasses, but will only check the growth of many coarse grasses (see Table 2). The highest dose will be required to give adequate control of most coarse grasses. Broadleaved weeds are not well controlled. Granular atrazine is less effective than wettable powder atrazine and should only be used for susceptible grasses, and preferably at the highest rate.

2.17.2.   All common forest conifers used in Britain tolerate the recommended rates with negligible damage, although Norway spruce, Western hemlock and European larch are more sensitive and should only be sprayed at the lowest dose. Little is known about the tolerance of broadleaved trees, although beech and oak have been sprayed without damage. In broadleaved crops atrazine should be applied before the trees come into leaf. Trees are more sensitive in the first season after planting and, if the weeds are susceptible to atrazine, it is advisable to use only the lowest doses. Atrazine is also more effective and more likely to damage the crop on soils with a low organic matter content in the upper horizons (say less than 7%—e.g. recently cultivated soils), and only the lowest recommended doses should be used in these situations.

Atrazine is *not effective* on organic soils (peats).

2.17.3.   Trees do not need to be protected from either formulation of atrazine. The wettable powder can be applied from knapsack sprayers at low pressure using fan jets. (*N.B.* care must be taken to ensure that the powder is properly mixed and dispersed throughout the water and that it does not settle out subsequently). Bulk mixing of large quantities in tanks is not feasible unless the tank incorporates a system to keep the contents continuously agitated. Granules can be applied as for chlorthiamid or dichlobenil (see para 2.18.5).

2.18. CHLORTHIAMID as 40 to 60 kg 7½% w/w granules per hectare (3·0 to 4·5 kg a.i./ha) applied from January to March in southern Britain, or from January to April in northern Britain, only in crops of tolerant species.

2.18.1.   The lower dose is suitable for light weed growth, or when trees are tall enough to require less than complete weed control for release. Also, since control tends to be better at the later dates of application, lower doses can be considered for March–April applications. Chlorthiamid is effective against a wide range of grass and broadleaved weeds and is a very useful herbicide for mixed weed situations. As with dichlobenil (para 2.20) it will sometimes give useful control of bracken, although it is too erratic in this respect to be used solely for bracken control.

2.18.2.   Chlorthiamid (and dichlobenil) are less effective on highly organic soils.

2.18.3.   Crops of Sitka spruce and Norway spruce, Corsican, Lodgepole and Scots pine, oak, ash, beech and sycamore, tolerate chlorthiamid as recommended above, although occasional trees may become chlorotic or even die. Douglas fir, Western hemlock, larches and *Abies* species should not be treated. There is insufficient evidence on the tolerance of other species to make recommendations.

2.18.4.   The yellowing of chlorthiamid treated trees appears to have little effect on their growth or survival, but discolouration may be of economic importance in Christmas tree plantations. Deaths are usually associated with bark lesions or "patches of dead bark" around the root collar zone.

2.18.5.   Applications of chlorthiamid are best made with mistblowers designed or modified to blow granules instead of liquids (see paras 8.27–8.30) although a satisfactory knapsack gravity feed applicator developed for dichlobenil by Duphar-Midox, Ltd may also be suitable for chlorthiamid. Application can be made by hand using a "sugar shaker" type container, but it is thought that much of the damage in early trials was due to uneven hand applications giving high concentrations of chlorthiamid around the root collar of trees. Since chlorthiamid is quite expensive, it is usual to apply it only to strips along the rows of trees or to patches round each tree.

2.19. DALAPON as 13·5 kg of 74% w/w commercial product per hectare (10 kg a.e./ha) in water at medium volume to grass weeds whilst they are growing reasonably vigorously, and whilst the crop is dormant.

2.19.1.   As grasses need to be growing actively to get a satisfactory kill, the best time of application is in March or early April, just before the tree begins to grow. Applications can also be made in October/November, but the weeds frequently recover before the end of the following growing season. Applications should not be made immediately after rain or if rain is anticipated, as the herbicide is easily washed from sprayed foliage.

2.19.2.   Dalapon has little effect on broadleaved weeds, and should only be used where grasses are the main problem. It is probably the most effective herbicide for rhizomatous grasses and *Calamagrostis* species although the latter is still not completely killed.

2.19.3.   Dalapon is effective on most soil types because it is mainly absorbed through weed foliage.

2.19.4.   The spray should be directed so as to minimise wetting the crop foliage. This may be done by using a lance fitted with a short boom having a jet at each end, the boom being held over the row of trees so that the jets spray a strip about 1·5 metres wide on either side of the row. Alternatively, the tree can be protected from the spray by a guard (e.g. a "Politec").

2.20.   DICHLOBENIL as 40 to 60 kg 7½% w/w granules per hectare (3·0 to 4·5 kg a.i./ha) applied from January to March in southern Britain or from January to April in northern Britain, only in crops of tolerant species.

2.20.1.   Dichlobenil should be used in exactly the same way as chlorthiamid (para 2.18) all the information under chlorthiamid being relevant.

2.20.2.   Dichlobenil is a closely related compound to chlorthiamid, the latter breaking down to dichlobenil in the soil shortly after application. However, dichlobenil is less soluble in water and more volatile than chlorthiamid, and these properties may make it less reliable than chlorthiamid in drier areas, although the latest granular formulations of dichlobenil appear to overcome this problem.

2.21.   PARAQUAT as 5 litres of 20% w/v concentrate per hectare (1·0 kg a.i./ha) in water at medium volume to green grass and herbaceous broadleaved weeds, avoiding all contact with the crop trees.

2.21.1.   There are two suitable periods for applying paraquat:

(1) In the spring and early summer to short green grasses and herbs. Foliage present at the time will be killed quickly, but regrowth is often rapid; two months later there may be a complete vegetation cover, but this will not be so tall as unsprayed vegetation. Spraying early in the season when the vegetation is not tall enough for hand weeding allows the weeding season to be extended—a distinct advantage when the labour force is limited.

(2) Autumn or late winter whilst the vegetation is still green. (Autumn applications usually give best kill). The kill or check is often much greater than the spring and early summer treatment and control may persist the whole of the next growing season. Some form of shield must still be used and the vegetation must be short. Ideally, this treatment should be used following hand weeding, where the vegetation has grown through the cut material; or on areas sprayed with paraquat in the spring and early summer.

2.21.2.   Paraquat is more effective on grasses than broadleaved weeds, but most broadleaved weeds present at the time of spraying will be severely checked. Living, aerial parts of broadleaved weeds are rarely present if spraying is carried out in early autumn or late winter, so spraying at this time is not so useful where broadleaved weeds are expected to be a problem.

2.21.3. Do not spray paraquat onto dead or cut foliage, or foliage which is severely browned (e.g. in drought conditions) as it will be ineffective. However, paraquat is extremely "rainfast" and may be used on dry or slightly damp foliage irrespective of the expected weather conditions.

2.21.4. Paraquat is effective on all soil types, including peat soils, because it relies entirely on foliar contact and absorption for its activity.

2.21.5. Paraquat must not be applied to the crop trees, especially to the foliage. Even *stems* of very young trees are susceptible to serious damage. The best method of ensuring adequate crop protection is to use one of the guards or cones designed for the purpose. Paraquat applied to tall foliage may be transferred to nearby crop trees as the foliage springs back after removing the sprayer. Tall or heavy weed growth also makes the use of a shield difficult.

2.22. (Information) Glyphosate

Glyphosate applied as 7 to 13 litres of 36% w/v concentrate per hectare (2.· to 5·0 kg a.e./ha) in March or June in medium volume has shown considerable promise for controlling grass and herbaceous broad-leaved weeds in young plantations.

2.22.1. Early indications are that this herbicide is very effective against all grasses, and that it will severely check or kill many broadleaved weeds, provided there is sufficient live leaf area present at the time of spraying. It appears to be extremely well translocated within the plant and to give, therefore, such a complete kill of sprayed weeds that control lasts for the whole season.

2.22.2. Crop trees are likely to be damaged by overall sprays of glyphosate. They should not be damaged by applications placed to miss their foliage.

2.22.3. It is anticipated that this very promising herbicide will be most useful for controlling mixtures of grass and herbaceous broadleaved weeds—maybe where woody weeds (e.g. bramble) are also a problem.

2.23. (Information) Propyzamide

Propyzamide applied as 4 to 8 kg 50% w/w wettable powder or as 50 to 100 kg 4% w/w granules per hectare (2 to 4 kg a.i./ha) in water at medium volume has recently shown promise for controlling mainly grass weeds in young plantations, particularly coarse grasses like *Deschampsia caespitosa*.

2.23.1. The optimum time for application appears to fall between October and January. Corsican pine, Scots pine, Japanese larch, Norway spruce, Sitka spruce, Red cedar, beech and oak planted in early November and subsequently sprayed in mid-November or in mid-January have been undamaged. This early winter application time could be of particular value for using labour at this less busy time of the year.

CHAPTER 3

RECOMMENDATIONS II: WOODY BROADLEAVED WEEDS AND MIXTURES OF WOODY AND HERBACEOUS BROADLEAVED WEEDS (OTHER THAN HEATHER AND RHODODENDRON)

GENERAL CONSIDERATIONS

3.1.   Table 3 sets out the relative susceptibility of the common woody weed species to 2,4–D, 2,4,5–T and ammonium sulphamate by the normal methods of application. Most species are best controlled with 2,4,5–T, taking into account both general effectiveness and cost. Ammonium sulphamate should be used for those species which are not well controlled by 2,4,5–T; 2,4–D is used for one or two special purposes. Other herbicides are known to be effective against woody weeds. 2,3,6–TBA (Holmes, 1957), 2,4,5–TP Holmes, 1963) picloram (Aldhous, 1965), and picloram and cacodylic acid (Brown and Mackenzie, 1971), have all been tested to some extent in Britain for controlling woody broadleaved weeds. However, none of these has been considered to be so generally effective or so safe to the crop, user and environment as 2,4–D, 2,4,5–T or ammonium sulphamate.

3.2.   When considering which herbicide technique to use, it should be remembered that large standing dead trees are unsightly, especially if in an isolated or prominent position. Even small crops, so killed, can look extremely conspicuous when viewed from a neighbouring hill slope.

3.3.   Coppice and scrub trees, if killed standing, disintegrate slowly and do no damage to an undercrop. On the other hand, falling debris from large trees may constitute a danger to passers-by. Land owners or tenants who have used herbicides, or who have arranged their use to kill trees growing near public roads or footpaths, may be liable for injury to passers-by or damage to their property from falling branches etc. Large trees near roads and footpaths should, therefore, never be killed standing.

3.4.   The use of esters of 2,4,5–T and 2,4–D with a volatility higher than the iso-octyl esters should be avoided. Esters of moderate or high volatility can volatilise from treated surfaces shortly after application and the resultant vapour cloud may drift long distances and damage nearby sensitive crops (see paras 11.33, 11.34 and 12.36).

3.5.   In all stem and stump spraying operations, it may be necessary to add to the spray solution a marker dye to indicate which stumps have been sprayed, to check that the right stumps have been treated, both at the time and within three or four days of spraying. Suitable spray markers are listed in paras 12.38 and 12.39.

FOLIAGE SPRAYING OF BROADLEAVED WEEDS

3.6.   Foliage sprays should be applied evenly to all parts of the weed to be killed—foliage, branches and stem down to soil level. Leaves will obviously receive by far the greatest amount of spray, but, especially with more resistant species, the aim should be to cover all parts of the plant.

3.7.   Because woody weeds are often unevenly distributed, foliage sprays are not always overall sprays (see Glossary), but may often only need to be applied to spots or patches of weed.

TABLE 3

THE EFFECT OF 2,4,5-T, 2,4-D AND AMMONIUM SULPHAMATE ON COMMON WOODY WEEDS

| WEED SPECIES | 2,4,5-T | | | | 2,4-D | | Ammonium Sulphamate |
| | Method of Application | | | | Method of Application | | Method of Application |
	Foliage	Basal bark	Stumps	Injections	Foliage	Injections	Frills, notches or stumps
ALDER (*Alnus glandulosa*)	S	MS	S		MS		S
ASH (*Fraxinus excelsior*)	MR	MR	MS	MR	MR	MR	S
BILBERRY (*Vaccinium* spp)					MS		
BEECH (*Fagus sylvatica*)	MS	MS					
BIRCH (*Betula* spp)	S	S	S	MS	MS	MS	S
BLACKBERRY (*Rubus* spp)	S		S		MR		
BLACKTHORN (*Prunus spinosa*)	S	MS	S				S
BOX (*Buxus sempervirens*)	MS						
BRIAR (*Rosa* spp)	S		S		MR		
BROOM (*Sarothamnus scoparius*)	S		MS		S		
BUCKTHORN (*Rhamnus cathartica*)	MS	MS					
HORSE CHESTNUT (*Aesculus hippocastanum*)	MS	S	S				
SPANISH CHESTNUT (*Castanea sativa*)	MS	S	S			MS	S
DOGWOOD (*Thelycrania sanguinea*)	S	MS	MS				
ELDER (*Sambucus nigra*)	S	S	S		MS		S
ELM (*Ulmus* spp)	MS	MS	S		MR		
GORSE (*Ulex* spp)	MS		S		MR		
HAWTHORN (*Crataegus* spp)	MR	MR	MR		R		S
HAZEL (*Corylus avellana*)	MS	MS	S		MR		
HEATHS (*Erica* Species 4)					MS		
HEATHER OR LING (4) (*Calluna vulgaris*)	MR				MS		
HOLLY (*Ilex aquifolium*)	R		MS		R		
HORNBEAM (*Carpinus betulus*)	MS	MS	S				
IVY (*Hedera helix*)	MR				R		
JUNIPER (*Juniperus communis*)	R				R		
LIME (*Tilia* spp)	MS	MR	MS	MR		MR	

TABLE 3—*continued*

WEED SPECIES	2,4,5-T Method of Application				2,4-D Method of Application		Ammonium Sulphamate Method of Application
	Foliage	Basal bark	Stumps	Injections	Foliage	Injections	Frills, notches or stumps
FIELD MAPLE (*Acer campestre*)	MS	S	S				
OAK (*Quercus* spp)	MR	MS	MS	MS	MR	MS	
WILD PEAR (*Pyrus communis*)	MS		S				
POPLAR (*Populus* spp)	S	S	S		MS		
PRIVET (*Ligustrum vulgare*)	MS						
RHODODENDRON (5) (*Rhododendron ponticum*)	MR	MR	MR		R		MS
ROWAN (*Sorbus aucuparia*)	MS	MS	MS		MS		
SNOWBERRY (*Symphoricarpus* spp)	MR	MR	MR		S		
SYCAMORE (*Acer pseudoplatanus*)	MS	MR	S	MR			MS
WILLOW (*Salix* spp)	S	MS	MS		S		S
(5) LAUREL (*Prunus laurocerasus*)	MR						MS

NOTES TO TABLE 3

1. Definition of weed susceptibility

   S =Susceptible: Consistently good control by suggested technique at the lower of the application doses recommended.

   MS =Moderately Susceptible: Good control with the higher of the doses recommended.

   MR=Moderately Resistant: Some effect from the higher doses of applications but recovery rapid.

   R =Resistant: No useful effect at the highest dose quoted.

2. Growth-regulating herbicides are most concisely prescribed in terms of the acid, whichever derivative is used and however this is formulated (e.g. 2,4-dichlorophenoxyacetic acid as the sodium or potassium salt, amine, emulsifiable ester, etc). Full details of the recommended formulations and their doses are given in the text.

3. Recommended doses of application are:—

   2,4,5-T ⎤ Overall summer foliage sprays: 2–4 kg acid per hectare.

   and   ⎬ Basal bark and cut-stump treatments: 1·5–2·5 kg acid per 100 litres of paraffin applied as per text.

   2,4-D ⎦ Injection: 1·0 ml of 50% w/v 2,4,5-T or 2,4-D applied as per text.

   Ammonium sulphamate: 0·4 kg per litre of water applied as per text.

4. For the best means of controlling heather, see paras 4.1 to 4.7.

5. Rhododendron and laurel can be adequately controlled with higher rates of 2,4,5-T. See paras 6.1 to 6.5.

3.8.   Foliage sprays should not be applied at volumes which cause run-off, as this is wasteful of herbicide and it increases the cost. Applications should normally be made at low volume (LV) using a mistblower. Ultra low volume (ULV) applications have recently been tested and provisional recommendations for the use of this technique are made.

3.9.   Deciduous species must be sprayed when foliage is healthy and functional. Foliage sprays are not so effective when applied to leaves which are rapidly expanding at the beginning of the growing season (May to mid-June) and may be wasted entirely if applied shortly before leaf fall in the autumn.

3.10.   Where coppice shoots are to be sprayed, treatment should be delayed until a substantial area of leaf surface has developed, especially if the coppicing stumps are large. Coppice shoots should not normally be treated by foliar spraying (see Stump Treatments) until there is one or two years' growth present and a well developed leaf area.

3.11.   After foliage spraying, treated shoots must not be cut for at least a month to allow the herbicide to be fully translocated into the woody tissues.

FOLIAGE SPRAYING BEFORE PLANTING

3.12.   **2,4,5–T ESTER as 7 litres of 50% w/v 2,4,5–T emulsifiable concentrate per hectare (3·5 kg 2,4,5–T acid/ha) applied in water at low or medium volume for woody species shown as susceptible or moderately susceptible in Table 3, at any time between when leaves are fully expanded (June?) until late August/ September (before first signs of leaf (fall). Evergreen species can normally be sprayed all the year round.**

3.12.1.   Up to 20% of paraffin by volume added to the 2,4,5–T concentrate, before diluting with water, may increase the effectiveness on more resistant species. The oil must be well mixed with the concentrate before diluting with water.

3.12.2.   If the woody weeds present a physical obstacle to proposed ground preparation and planting operations (e.g. bramble, gorse, etc), then spraying should take place sufficiently in advance of these operations to allow the weeds to die back or collapse. Many species take 1–2 years to die back fully or collapse. Access lanes may have to be cut at the time of spraying, in which case cut stumps and stems in the lanes should also be treated to prevent regrowth (see paras 3.29–3.36).

3.13.   **2,4,5–T/2,4–D ESTER MIXTURE as 4·5 litres of 25% w/v 2,4,5–T plus 50% w/v 2,4–D emulsifiable concentrate per hectare (3·5 kg total acid/ha) applied in water at low or medium volume for mixtures of woody and herbaceous broadleaved weeds, or even for herbaceous broadleaved weeds alone, during the same period as recommended for 2,4,5–T ester alone (see para 3.12); if a herbaceous broadleaved weed is the major problem, timing may have to be adjusted to fit its development.**

3.13.1.   This mixture should only be used if the herbaceous broadleaved weed component is a problem. Woody weeds will still take 1–2 years to die back fully or collapse.

3.14.   **2,4,5–T ESTER as 10 litres of a special ULV formulation containing 30% w/v 2,4,5–T acid in a non-phytotoxic oil (3·0 kg acid/ha) through a ULV sprayer. No dilution is required. The recommended time of application is as for normal low volume 2,4,5–T applications (see para 3.12).**

3.14.1.   Recent trials suggest that ULV application will provide similar control of woody species to that provided by mistblower applications, although species normally moderately resistant to 2,4,5–T applied by mistblower have sometimes exhibited greater resistance to 2,4,5–T applied by ULV sprayers.

3.14.2.   Operators need special training and should use special light-weight protective clothing to get the best out of ULV spraying, and this method should only be used if these requirements can be met at the outset.

3.14.3.   ULV applications of 2,4,5–T increase the area sprayed per man day and, although the special formulation of 2,4,5–T costs more than normal, 2,4,5–T formulations, the final cost per hectare may be less than with mistblower applications.

FOLIAGE SPRAYING AFTER PLANTING

(a) Overall Applications (in conifer crops only).

3.15.   **2,4,5–T ESTER as 4·5 to 7·0 litres of 50% w/v 2,4,5–T emulsifiable concentrate per hectare (2·25 to 3·50 kg acid/ha) applied in water at low or medium volume for woody species shown to be susceptible or moderately susceptible in Table 3, provided that spraying is confined to a period late in the summer which commences when the conifer crops has ceased shoot extension and formed resting buds, and before the leaves of the weeds are showing signs of autumn senescence. Spraying of evergreen weeds can continue throughout the autumn and winter until about 14 days before the conifer crop flushes in the spring.**

3.15.1.   For most crop/deciduous weed combinations the late summer spraying period is from about mid-August to early October in the south and west of the country, and from late-July to end of September in the north and east.

3.15.2.   However, conifer species vary in their tolerance of overall applications of 2,4,5–T. Spraying may take place at any time during the above periods in crops of Scots pine, Corsican pine, Norway spruce, Sitka spruce, Douglas fir, *Abies grandis* and *Abies procera*. Applications should be confined to the period from September/October to 14 days before flushing in crops of Western hemlock, Western red cedar and Lawson cypress and then only at the lowest dose recommended. Overall applications should not be made in crops of Lodgepole pine, *Pinus radiata*, *Pinus pinaster* and larches unless serious damage to these species can be accepted—as for instance when they form an unimportant part of a mixed crop (but see Placed Applications).

3.15.3.   Even crop species in which overall applications of 2,4,5–T are recommended may be slightly checked or damaged by the herbicide, needle browning or twisting (especially of lammas shoots) appearing as the symptoms. This type of damage can be reduced by directing the spray away from trees wherever possible (only applicable to ground spraying).

3.16.   **2,4,5–T/2,4–D ESTER MIXTURE as 3·0 to 4·5 litres of 25% 2,4,5–T plus 50% 2,4–D emulsifiable concentrate per hectare (2·25 to 3·38 total acid/ha) applied in water at low or medium volume for mixtures of woody and herbaceous broadleaved weeds, or even for herbaceous broadleaved weeds alone, during the same period as recommended for 2,4,5–T ester alone (see para 3·15).**

3.16.1.   This 2,4,5–T/2,4–D mixture should only be used if the herbaceous broadleaved weed component is a problem.

3.16.2.   Crop species tolerance of 2,4,5–T/2,4–D mixture is similar to their tolerance of 2,4,5–T alone, and the same restrictions apply (see para 3.15.2).

3.17.   **2,4,5–T ESTER as 7 litres of a special ULV formulation containing 30% w/v 2,4,5–T acid in a non-phytotoxic oil (2·1 kg acid/ha) through a ULV sprayer. No dilution is required. The restrictions on spraying period and crop species are the same as for normal low volume applications of 2,4,5–T ester (see para 3.15).**

3.17.1.   See para 3.14 also for comments on the reliability and advantages of this technique, and the need for special training and protective clothing.

(b) Directed Applications (to miss crop trees)

3.18.   2,4,5–T ESTER OR 2,4,5–T/2,4–D ESTER MIXTURES at the same rates as recommended in paras 3.15 and 3.16 may be used to control patches of woody weeds or woody/herbaceous broadleaved weeds in crops of species sensitive to overall applications of 2,4,5–T (conifers and broadleaves) in late summer, or in species which tolerate late summer overall applications of 2,4,5–T during the period from when the leaves of the weed species are fully expanded (June?) until late August/September (before the first signs of leaf fall), provided such applications can be directed so as to miss crop trees.

3.18.1.   Applications should only be made at medium volume, using a knapsack sprayer at low pressure ($< 1 \cdot 0$ kg/cm^2—15 pounds per square inch) and applications should not be made during the period when crop trees are growing most vigorously (May/early June).

STEM APPLICATIONS TO BROADLEAVED WOODY WEEDS

3.19.   Where coppice or scrub is considered unsaleable, it may be desirable to kill it standing rather than to clear it. In other situations, planting may have taken place before complete removal of the previous crop to take advantage of its overhead cover for giving shade and protection against frost; as the new crop develops and the cover has to be removed, it may be cheaper and otherwise preferable to kill it standing, rather than fell and risk damage to the young crop. The residual and slowly decreasing protection afforded by the dead stems can be of value silviculturally. However, consideration should always be given to the effect of standing, killed, woody vegetation on access to the new plantation for subsequent maintenance and to its effect on the landscape.

3.20.   There are three methods of applying herbicides to the stems of broadleaved trees to kill them standing; basal bark applications, applications to frills or notches cut in the stem, or injections directly into the transporting tissues of the tree. The method chosen depends on the average size of tree to be killed, the accessibility of each individual stem and the sensitivity of the weed tree to 2,4,5–T or 2,4–D.

BASAL BARK APPLICATIONS (2,4,5–T ONLY) Plate 3.

3.21.   2,4,5–T ESTER in paraffin, at a concentration of 2·0 to 3·0 litres of 70–80% unformulated low-volatile ester or 3·0 to 5·0 litres of 50% emulsifiable low-volatile ester per 100 litres of paraffin (1·4 to 2·4 kg acid/100 litres of paraffin), to the whole circumference of the bottom 30 to 45 cm of the stem.

3.21.1.   The application, which is normally made by spraying, should saturate the bark to the point of run-off. The object is for the herbicide to penetrate the bark and enter the tree's transportation tissues for translocation round the tree.

3.21.2.   To get adequate penetration it is essential to use a hydrocarbon oil as a diluent. Water gives insufficient penetration. For safety, premium grade paraffin conforming with Class C1 of British Standards Specification 2869 is recommended.

3.21.3.   Bark thickness influences penetration. When bark is thin, application to the bottom 30 cm of the stem is usually sufficient, but for thicker bark the bottom 45 cm should be treated. Trees over 10 cm diameter at breast height usually have bark which is too thick to allow adequate penetration, even by paraffin-borne, 2,4,5–T.

19

3.21.4.   The kill of species not very susceptible to 2,4,5–T (see Table 3) can be improved by spraying the greater length of the stem base.

3.21.5.   Basal bark applications can be used both before and after planting, provided direct contact between the herbicide and crop trees can be avoided. Applications are effective in all months, but it is advisable to avoid the growing season, particularly just before and during the period of most active growth (April/May/early June). Broadleaved crop species are more likely to sustain damage during the growing season than conifers.

3.21.6.   An additional reason for these restrictions is that 2,4,5–T may volatilise from treated surfaces shortly after application in periods of hot weather, and the drifting vapour damage the crop.

3.21.7.   About 1 litre of spray solution will be required for each 100 cm of stem diameter, or 100 litres for 1000 stems of 10 cm average diameter.

FRILL GIRDLING, NOTCHING AND TREE INJECTION METHODS
("CUT BARK" METHODS) Plates 4–6.

3.22.   Stems larger than 10 cm diameter at breast height are more effectively treated by frill girdling, notching or by tree injection, although the latter method may also be suitable for smaller stems. All are simply means of improving the penetration of the herbicide into the trees' translocating tissues. Water soluble herbicides like 2,4–D amine and ammonium sulphamate must be applied using a cut-bark method for all sizes of stem (ammonium sulphamate should not be used in a tree injector because of the corrosion it will cause.

3.23.   Cut-bark applications involve more careful placing of the herbicide than do basal-bark or stump treatments; there should therefore be little risk of spraying crop trees or of volatilisation from large areas of sprayed surface. Thus both pre- and post-planting applications are recommended, with no restriction on season of application. Applications in any month are effective.

3.24.   **2,4,5–T ESTER IN PARAFFIN at the same concentrations as recommended for basal bark treatments (see para 3.21) to freshly cut frills in the stems of standing unwanted trees, when these are over 10 cm diameter.**

3.24.1.   Frill girdling involves preparing a "frill" in each tree as near to ground level as is convenient just prior to application of the herbicide. This "frill" is a ring of downward-sloping overlapping cuts made with a light axe or bill-hook (see plate 5). The cuts must penetrate to the cambium, and, if possible into the outer sapwood. To complete the treatment, the herbicide solution is then sprayed onto the full circumference of the stem just above the frill so that it runs down into the cuts. With 2,4,5–T in paraffin, about 30 cm of stem above the frill may also be sprayed to improve the kill, especially on species resistant to 2,4,5–T.

3.24.2.   Applications are best made using a knapsack sprayer or live reel sprayer (e.g. "Pharos") whose lances are fitted with a solid-stream jet. However, small programmes can be completed with plastic watering cans fitted with simple spouts.

3.24.3.   One litre of spray solution will treat from about 100 to 500 cm of stem diameter (e.g. 100 litres will treat 500 to 2500 stems of 20 cm diameter—depending on how much of the bark above the frill is sprayed).

3.25.   **AMMONIUM SULPHAMATE SOLUTION at a concentration of 0·4 kg ammonium sulphamate per litre of water, to freshly cut frills in the stems of unwanted trees of all sizes.**

3.25.1.   Frills are cut as described in para 3.24.1 above. This treatment is particularly suitable for species resistant or moderately resistant to 2,4,5–T and 2,4–D (see Table 3). Otherwise 2,4,5–T and 2,4–D treatments are cheaper and more convenient.

3.25.2.   A plastic watering can fitted with a suitable spout is probably the best tool for applying ammonium sulphamate solution to frills. This solution is corrosive to most metals and quickly ruins sprayers in which metal parts are exposed to the herbicide (see para 6.4.3).

3.25.3.   Trees larger than 15 to 20 cm diameter at breast height are not so well controlled by applications of ammonium sulphamate to frills, probably because insufficient herbicide can be held in the frills to kill them. For large trees over 15 to 20 cm diameter at breast height which are resistant to 2,4,5–T, use the notch method of applying ammonium sulphamate crystals (see para 3.26).

3.26.   AMMONIUM SULPHAMATE DRY CRYSTALS to notches cut round the stem of trees over 15 to 20 cm breast height diameter at a rate of 15 g ammonium sulphamate per notch. Notches must be not more than 10 cm apart edge to edge.

3.26.1.   Notching involves cutting a ring of "steps" in the stem of the tree to be killed using an axe. The base of the step should slope slightly downwards to help retain the ammonium sulphamate crystals, and the notches should be no more than 10 cm apart edge to edge (see plate 6).

3.27.   2,4,5–T EMULSIFIABLE ESTER (0·5 kg acid/litre) injected undiluted at 1·0 ml per injection point at 7·5 cm centres round the circumference of species that are susceptible or moderately susceptible to 2,4,5–T.

3.27.1.   If emulsifiable ester (0·5 kg/litre) is not immediately available, a suitable injection solution can also be made up from unformulated esters of 2,4,5–T by diluting them with paraffin until they contain 0·5 kg acid/litre. Undiluted they tend to be too viscous to flow through the injector satisfactorily. There is some evidence that water soluble forms of 2,4,5–T (e.g. amines or potassium salt) are more effective than esters for tree injection, but these forms are not commercially available currently in Britain (1975).

3.27.2.   Injection is a method of inserting concentrated herbicides, including 2,4,5–T and 2,4–D (see para 3.28 below) directly into the translocating tissues of the tree.

3.27.3.   Some injection equipment makes both the incision and injection. Injection with some models is automatic as the blade hits the tree, and with others injection is made by operating a lever after the blade has been inserted. The apparent advantages of such equipment are, to some extent, offset by the somewhat clumsy nature of the equipment and the need for good accessibility to each stem. Recently, equipment which separates the incision making and injection process has been developed, (almost a return to frilling and spraying, except that the above concentrated herbicide solution is used), and this is less demanding on accessibility and cheaper. (See Chapter 8 "Equipment".) Whatever type of injector is used, it is important that the incision is at least 3 cm wide horizontally, and penetrates to the inner bark/outer sapwood. Injections made into small holes are not so effective.

3.28.   2,4–D AMINE (0·5 kg acid/litre) injected undiluted at 1·0 ml per injection point at 7·5 cm centres round the circumference of species susceptible or moderately susceptible to 2,4,5–T.

3.28.1.   This treatment appears to be almost as effective as injecting 2,4,5–T ester, and because 2,4–D amine is cheaper than 2,4,5–T ester it can be considered an acceptable alternative.

3.28.2.   There is some evidence that, once they have penetrated to the cambium region, water soluble forms of a herbicide are more effective than oil soluble forms, even if the latter have been made miscible with water by the addition of emulsifiers. This may explain why 2,4–D amine is effective, whilst 2,4–D ester is not.

STUMP APPLICATIONS TO BROADLEAVED WOODY WEEDS

3.29.   After felling, hardwood stumps can readily be treated to prevent growth of unwanted coppice-shoots. The treatment is easy to apply and can result in considerable savings in the cost and number of subsequent weedings; the cost of cleaning at the time of brashing and first thinning can also be reduced.

3.30.   Stump treatments are best applied before planting, soon after the standing trees have been felled. Alternatively, they may quite well be applied after planting, though more care will be required in placing sprays.

3.31.   As the volume of spray solution required is proportional to the surface area of the stump, the lower a stump is cut, the less the area of bark that has to be treated. Leaving high stumps can add 20% or more to the cost of spraying, and may also create difficulties for subsequent weeding or harvesting equipment.

3.32.   There is nothing that will effectively and quickly rot the stump once it has been killed. If stumps have to be removed, it is usually easier to remove the stem and stump out together (using big tractors or winches) and then to cut off the stump, rather than to fell the tree and be faced with the task of lifting out the stump. It has been claimed that it is easier to remove a stump that has been killed with 2,4,5–T or ammonium sulphamate, than stumps which have some live roots.

3.33.   There are machines which will chip away stumps to about 25 cm below ground level. These machines are for hire in several parts of the country.

3.34.   **2,4,5–T ESTER IN PARAFFIN at a concentration of 2·0 to 3·0 litres of 70–80% unformulated low volatile ester or 3·0 to 5·0 litres of 50% emulsifiable low volatile ester per 100 litres of paraffin (1·4 to 2·4 kg acid/100 litres of paraffin) to the surface and bark of the stump.** Plate 7.

3.34.1.   The herbicide should be applied to saturate both the newly cut surface of the stump and to the surface of the remaining bark between the cut and the ground. Control of stumps over 25 cm in diameter can be improved by putting a frill or notches into the remaining bark before spraying.

3.34.2.   Ideally an application should be made immediately after felling and certainly no more than one week after felling. The kill of older stumps can be improved by letting them produce coppice shoots and then spraying both the bark and base of the young coppice shoots as well as the top and bark of the stump, or allowing sufficient foliage to develop to allow them to be treated with a foliar spray (see para 3.12–3.18). Cut stump treated areas should always be visited in the second growing season after treatment to "spot" treat any surviving stumps which reveal themselves by producing coppice shoots.

3.34.3.   2,4,5–T in paraffin can be applied at any time of the year, although it is often easier to apply from January to March/April because impeding vegetation is then at its lowest. The period late April to early June should be avoided for post-planting applications because even if direct spraying of crop trees can be avoided, warm weather is liable to cause volatilisation of 2,4,5–T from treated surfaces, and the vapour may drift onto crop trees. It is also less pleasant to spray when the weather is very hot. Stumps should not be treated when wet as this reduces the effectiveness of the treatment.

3.34.4.   Knapsack sprayers or live reel sprayers, operated at low pressure (> 1 kg/cm^2 = 15 pounds per square inch), and fitted with solid stream jets, are the best equipment for making cut-stump applications (see Equipment Section). About 1 litre of spray solution will be required for each 100 cm of stump diameter, or 100 litres for 500 stumps of 20 cm average diameter.

3.34.5. With stump sprays, it is often very useful to be able to trace stumps which have been treated. A marker dye such as Waxoline Red OS (oil soluble) may be used (see para 12.38). Dyes usually fade after a few days, so that checks should always be carried out shortly after spraying.

3.35. AMMONIUM SULPHAMATE SOLUTION at a concentration of 0·4 kg ammonium sulphamate per litre of water, to freshly cut stumps.

3.35.1. Application of ammonium sulphamate solution is made in the same way as described for 2,4,5–T above, except that application is concentrated on cut surfaces. The control of large stumps (> 25 cm diameter) can again be improved by putting frills or notches into the remaining bark.

3.35.2. For satisfactory control not more than two days must elapse between cutting and application. If the interval is increased beyond two days, control becomes less certain.

3.35.3. Applications are effective when applied in any season, although it is often easier to make applications from January to March/April when impeding vegetation is at its lowest.

3.35.4. Post-planting applications can be made, but there is a much greater risk of crop damage with ammonium sulphamate than with 2,4,5–T. This is because run-off and wash-off from treated stumps can contaminate surrounding soil, and trees rooting in this soil may be killed.

3.35.5. Applications can be made by knapsack sprayer in the same way as for 2,4,5–T, but the sprayer should be so constructed that the spray solution does not come in contact with the metal parts of the sprayer as these will then be rapidly corroded (see para 6.4.3). Otherwise, applications can be made from a plastic watering can.

3.36. AMMONIUM SULPHAMATE CRYSTALS at 6 g per cm of stump diameter to the freshly cut surface of stumps.

3.36.1. To improve the retention of the crystals the stump should be cut horizontal or preferably slightly V-shaped. Ammonium sulphamate should not be applied if heavy rain is expected within a few hours.

CHAPTER 4

RECOMMENDATIONS III: HEATHER

GENERAL CONSIDERATIONS

4.1.   Heather and heaths (mainly *Calluna vulgaris*, but with some *Erica* species) are common, dominant weeds in forest plantations on many infertile, upland soils in Britain, and on lowland heaths of southern Britain.

4.2.   Heather competition depresses crop growth, especially crops of Sitka spruce (*Picea sitchensis*) more than any other weed in British forestry. The mechanism of this competition has been fully discussed by Handley (1963) and it is sufficient here to note that a dense cover of heather causes nitrogen deficiency and a tendency towards phosphate deficiency, which is not always easily corrected until the heather is controlled. In spruce crops this deficiency can be so severe that the crop almost ceases growing.

4.3.   Situations in which spraying is essential or when fertilisation is the correct remedial measure are more fully discussed elsewhere by Wallis (in F.C. Leaflet 64, in press) and Everard (1974). However, Figure 2 summarises their recommendations. No recommendations are made for controlling heather in broadleaved crops because broadleaved species have only been rarely planted in situations where heather is a problem.

HEATHER CONTROL BEFORE PLANTING

4.4.   Experiments have not yet shown that killing heather before planting gives any worthwhile benefits, either in crop growth or in reducing the need for post-planting weeding. Previously unplanted areas are normally ploughed before planting and, *initially*, adequate suppression of heather is given by this operation. (If control is required to facilitate ploughing, this is more cheaply achieved by burning the heather). On sites to be replanted, the previous crops has usually effectively shaded out the heather. However, for the few instances when control may be required, recommendations are given below.

4.5.   **2,4,–D ESTER as 12 litres of 50% w/v emulsifiable concentrate per hectare (6 kg acid/ha) in water at medium or low volume, or as 15 litres of undiluted 40% w/v special Ultra Low Volume formulation per hectare (6 kg acid/ha). In either case, applications should be made from June to August (inclusive).**

4.5.1.   Application should be made to lanes about two metres wide either using a knapsack or mistblower for the medium or low volume applications, or the special Ultra Low Volume application equipment. It is important to get good, even coverage of the spray on the heather foliage and stems for satisfactory control. See the post-planting application section below for further details.

HEATHER CONTROL AFTER PLANTING

4.6.   **2,4–D ESTER as 8 litres of 50% w/v emulsifiable concentrate per hectare (4·0 kg acid/ha) in water at medium or low volume, or as 10 litres of undiluted 40% w/v special ULV formulation per hectare (4·0 kg acid/ha). In either case, the special techniques and times of application described below must be used for good control.**

4.6.1.   Single lane application (i.e. between two rows of trees) is recommended for all types of application as this is the best method of ensuring uniform, overall coverage of heather foliage and stems. Medium

24

WHEN TO CONTROL HEATHER IN YOUNG CONIFER PLANTATIONS

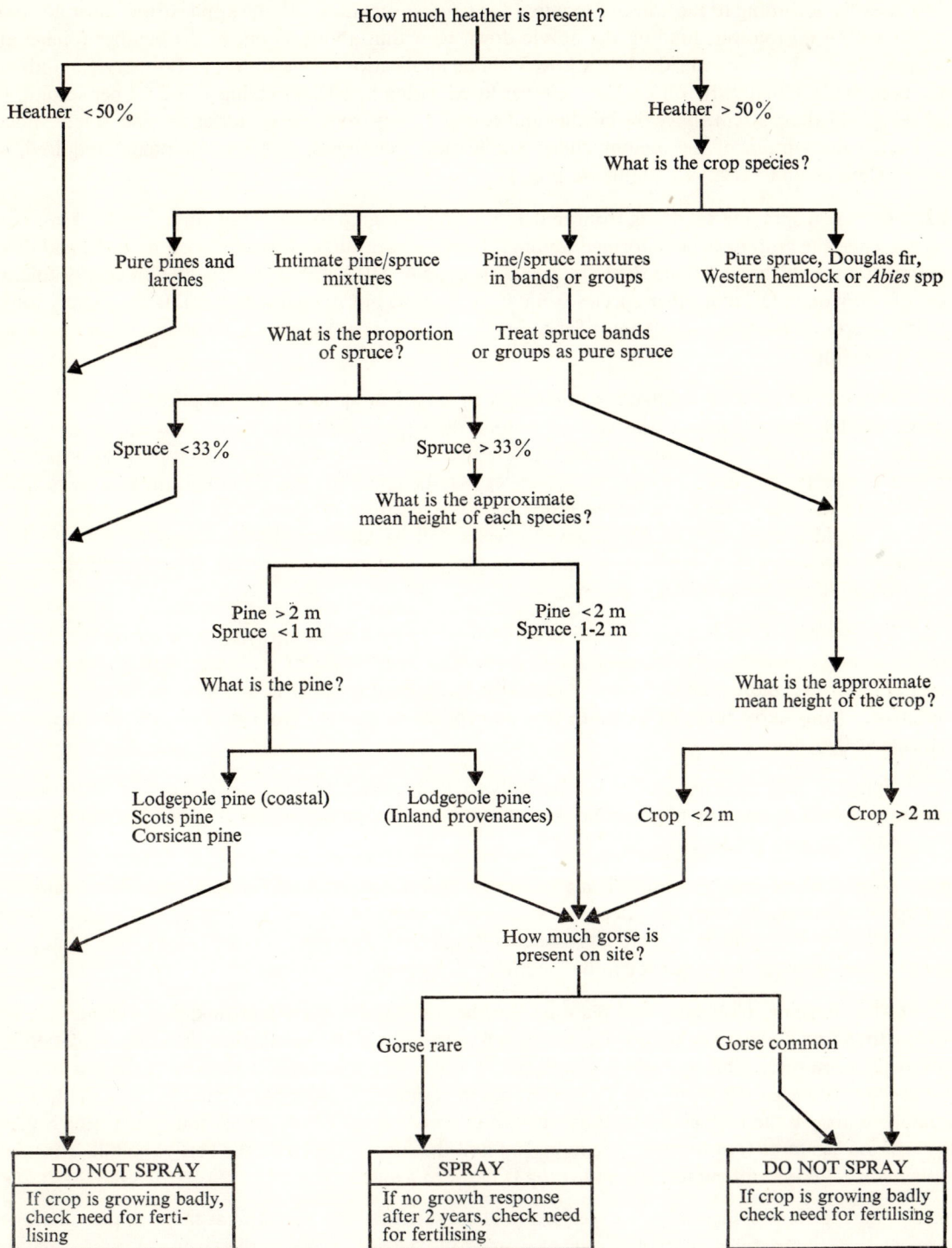

How much heather is present?

Heather < 50%

Heather > 50%

What is the crop species?

Pure pines and larches

Intimate pine/spruce mixtures

What is the proportion of spruce?

Pine/spruce mixtures in bands or groups

Treat spruce bands or groups as pure spruce

Pure spruce, Douglas fir, Western hemlock or *Abies* spp

Spruce < 33%

Spruce > 33%

What is the approximate mean height of each species?

Pine > 2 m
Spruce < 1 m

What is the pine?

Pine < 2 m
Spruce 1-2 m

What is the approximate mean height of the crop?

Lodgepole pine (coastal)
Scots pine
Corsican pine

Lodgepole pine (Inland provenances)

Crop < 2 m

Crop > 2 m

How much gorse is present on site?

Gorse rare

Gorse common

DO NOT SPRAY
If crop is growing badly, check need for fertilising

SPRAY
If no growth response after 2 years, check need for fertilising

DO NOT SPRAY
If crop is growing badly check need for fertilising

Figure 2. Algorithm to help decide when to control heather in young conifer plantations.

25

volume applications should be made with a knapsack sprayer and lance fitted with a red, blue or yellow politip fan jet, this range of jets allowing the volume of application to be adjusted from about 450 litres/ha to 110 litres/ha according to the amount required for good cover. Low volume applications must normally be made with a mistblower, holding the nozzle down to within about 30 cm of the heather foliage and waving it slightly from side to side. Ultra Low Volume applications have so far, in Forestry Commission trials, been made with a "Micron ULVA" sprayer fitted with a nozzle delivering 1 or 2 ml per second, the head being held about 30 cm above the heather and swung slightly from side to side as the operator advances. For a fuller description of the recommended application techniques, and the equipment required, see Forestry Commission Leaflet 64 Wallis (in press).

4.6.2.   The main spraying season is from mid-July or mid-August to early September, this season commencing when the crop trees have formed terminal buds and are sufficiently hardened to withstand direct contamination with the herbicide drift. In northern Britain, Sitka spruce is usually sufficiently tolerant from mid-July onwards, and other species from early August, but in southern Britain, the slightly longer growing season means that early August for Sitka spruce and mid-August for other species is a safe time to start spraying.

4.6.3.   Deposits on the crop foliage are minimised by inter-row spraying, and applications can be made in most conifer species because the risk of serious damage is slight; however, damage to Lodgepole pine, *Pinus radiata* and larches from LV and ULV applications does occur, especially if the crop is generally less than one metre tall, and in some circumstances may be unacceptable. Although there is no recorded experience of the effects on broadleaved crops following heather spraying with 2,4–D experience from elsewhere would suggest that all broadleaved species will be unacceptably damaged by LV and ULV applications. However, MV applications carefully placed to miss the crop may be a feasible technique of controlling heather in broadleaved crops.

4.6.4.   Forests with large heather control programmes can find extra spraying time by using knapsack sprayers to make placed applications of 2,4–D from May to mid-July or mid-August provided the mean height of the crop is more than one metre. Generally, however, it is better to avoid the period of maximum crop growth (May/early June) as trees are very susceptible to damage from drift or volatilisation of the herbicide at that time.

4.6.5.   Spraying after early September is often only partially successful because the heather becomes more resistant to 2,4–D. If spraying programmes have to be completed in September it is worth while increasing the dose to 5 kg acid per hectare.

4.6.6.   Bright, warm, sunny weather is an advantage but is not essential for spraying. Heavy rain soon after application may decrease the effectiveness of water-borne 2,4–D applications by knapsack and mistblower. ULV applications (in oil) are less likely to be affected. Spraying in wind speeds over 12 m.p.h. is not advisable, although a knapsack can be used in rather windier conditions.

4.6.7.   After burning, ploughing and planting, heather does not usually regrow quickly enough to be a problem for about 4 to 5 years. Crops younger than 4 years old are also usually less than one metre tall, and are usually more susceptible to herbicide damage. However, if the heather is allowed to get old and well-established it is more difficult to kill. The difference in response is probably related both to the leaf area available to absorb the herbicide and to the length of woody stem between leaf and root through which the herbicide has to be translocated. The more leaf and the shorter the stem, the better the kill. Therefore, heather control is best attempted when the crop is between 4 and 6 years old.

4.6.8.   There is some evidence from experiments in southern Britian that a phosphate or phosphate and nitrogenous fertiliser applied one or two seasons prior to spraying increases the vigour of old heather and its susceptibility to 2,4–D. This only has practical value if the plantation in question requires a top-dressing of fertiliser.

HEATHER CONTROL AND BEES

4.7.   Bees are attracted to heather when in flower. Ideally, for the bees, herbicides should *not* be applied when the heather is in flower (July to early September). In practice, this is usually the best season for treating the heather, and it may be impossible to avoid these months in forests with large spraying programmes. In such circumstances, any beekeepers with hives on or near the area to be treated must be notified well beforehand. Bees will normally avoid sprayed areas. Nevertheless, 2,4–D can cause mortality in bees if large areas are sprayed at one time and bees in the locality have no alternative nectar sources.

CHAPTER 5

RECOMMENDATIONS IV: BRACKEN

GENERAL CONSIDERATIONS

5.1.   Bracken is a major weed problem in plantations in many of the more fertile upland areas of Britain, and it is also a frequent weed in many lowland woods. However, in some areas bracken is considered a lesser evil than the weeds (e.g. grasses, bramble) that might invade after it has been effectively controlled. Certainly bracken is cheaper to cut by hand than many other weed types, although it may not be cheaper to cut by hand than to control with a herbicide. Also, any invading vegetation can now be effectively and cheaply controlled with other herbicides.

5.2.   Many areas with bracken also carry an understorey of grass. The development of a complete bracken cover often suppresses the grass growth, and the trees appear to grow successfully under the shade of the bracken fronds until later in the growing season than they would if surrounded and over-topped by heavy grass. Whilst the effect of such heavy bracken shading on the growth of the crop has not been investigated, it is true to say that the forester can afford to delay weeding in such areas without running a serious risk of crop losses. In a few areas (e.g. Thetford) the early bracken fronds may protect trees from late frosts.

5.3.   In spite of all these considerations, bracken must be weeded in young crops because if it falls on the crop trees after senescence, very high losses are experienced through "smothering". The latest herbicide treatments are now very competitive with hand weeding methods.

BRACKEN CONTROL BEFORE PLANTING

5.4.   **ASULAM as 7 to 10 litres of 40% w/v concentrate per hectare (2·8 to 4 kg a.e./ha) in water at low or medium volume after most of the bracken fronds have fully expanded (late June/early July) to well before any sign of senescence (late August).**

5.4.1.   Better control has been reported with ultra low and low volume applications than with medium volume applications, possibly because more asulam is retained on the frond with low volume applications. However, good even coverage of the fronds is the first consideration, and the volume of application should not be reduced at the expense of this.

5.4.2.   The optimum time for application is late June and July; control from August applications is, however, quite satisfactory, but the highest rates should be used. Asulam has little effect on the bracken cover in the year of application, but usually at least 70–80% less bracken will appear in the subsequent two growing seasons.

5.4.3.   **(Provisional) Asulam** at Ultra Low Volume.

Asulam as 7 litres of 40% w/v concentrate per hectare (2·8 kg ae/ha) at ultra low volume after most of the bracken fronds have fully expanded (late June/early July) to well before any sign of senescence (late August).

5.5.   **DICAMBA as 8 to 10 litres of 40% w/v concentrate per hectare (3·2 to 4·0 kg a.e./ha) in water at low or medium volume to within 4 months of planting.**

28

5.1.1.   The most effective time of application is one month either side of frond emergence, but application at any time of the year should give good control. Applications before the fronds emerge are attractive because of the ease of spraying. Application after full development of the fronds has little effect in the same season.

5.5.2. Dicamba is an expensive herbicide. Adequate control on moist, fertile soils seems to persist for only one to two years, but on dry, infertile sites adequate control often persists for four years or more. Clearly, it is more worthwhile to use dicamba on bracken growing on the latter site types.

5.6.   **(Provisional) M.C.P.A.** as 100 to 150 litres of 30% w/v potassium salt concentrate per hectare (30 to 45 kg a.e./ha) applied in water at medium volume during winter months has provided good initial control of bracken without damage to conifer crops planted in the following March or April.

5.6.1.   There is insufficient experience with M.C.P.A. to support a positive recommendation. Reports from other sources (Erskine, 1968; Martin, 1968) give conflicting results.

BRACKEN CONTROL AFTER PLANTING

5.7.   **ASULAM as 5 to 10 litres of 40% w/v concentrate per hectare (2·0 to 4·0 kg a.e./ha) in water at low or medium volume after most of the bracken fronds have fully expanded (late June/early July) to well before any sign of senescence (late August).**

5.7.1.   Better control has been reported with ultra low or low volume than with medium volume applications (see para 5.41) provided good even cover of the fronds is obtained. Better and more reliable control is obtained with 8 to 10 litres of concentrate per hectare, but 5 to 7 litres will normally provide adequate control, provided application is at the optimum time (late June and July).

5.7.2.   Asulam applied to fully expanded fronds has little effect in the year of application, and therefore it may still be necessary to cut dense bracken in this first year to protect the crop. At least 28 days must be left between application and cutting to ensure that the herbicide has been translocated to the rhizomes. Bracken cover should be reduced by at least 80% in the following two seasons.

5.7.3.   Corsican and Scots pine, Norway and Sitka spruce, Douglas fir, Grand fir, Japanese larch, beech, birch and elm of the common forest species show good tolerance of asulam at the above doses, although some might exhibit slight chlorosis and check at the highest rates and earliest dates of application. Western hemlock is more sensitive, and should be sprayed at not more than 7 litres of 40% concentrate only in August/early September. In normal field spraying the bracken canopy prevents most of the spray reaching the trees and so, in practice, the trees are generally safer than indicated.

5.7.4.   **(Provisional) Asulam at Ultra Low Volume.**

Asulam as 5 litres of 40% w/v concentrate per hectare (2·0 kg a.e./ha) at ultra low volume after most of the bracken fronds have fully expanded (late June/early July) to well before any sign of senescence (late August).

CHAPTER 6

RECOMMENDATIONS V: RHODODENDRON

GENERAL CONSIDERATIONS

6.1. On acid soils in the milder areas of the country *Rhododendron ponticum* can be a serious local problem. Replanting areas heavily infested with rhododendron is expensive, principally because they have to be cleared by hand or machine to allow planting, and rapid regrowth or re-invasion often necessitates expensive weeding operations during the early years of the crop.

6.2. There is no practical herbicide treatment that can be recommended for large, well established rhododendron bushes. However, once cut, regrowth from cut stumps or reinvasion by seedling rhododendron can be economically controlled with herbicides. For forests with a rhododendron problem, control with herbicides at the time of replanting should be standard practice if it is hoped to keep the rhododendron under control.

STUMP AND LOW REGROWTH CONTROL BEFORE PLANTING

6.3. **2,4,5–T ESTER in paraffin or water, at a concentration of 2·5 to 3·2 litres of 78% unformulated low-volatile ester (for paraffin only) or 4·0 to 5·0 litres of 50% emulsifiable 2,4,5–T in 100 litres of diluent (2·0 to 2·5 kg acid/100 litres of diluent) to all accessible surfaces, at any time of the year.**

6.3.1. For stump treatment 2,4,5–T in paraffin should be preferred. Applications should saturate to point of run-off all cut surfaces and bark down to soil level, and should ideally be made immediately after cutting or at least within one week of cutting.

6.3.2. Coppice shoots and seedling rhododendron can also be effectively and economically controlled provided they are less than about 1·5 metre high. The spray solution, using either paraffin or water as a diluent, should be applied to thoroughly wet all aerial parts (foliage, stems and stumps) of the bush. The volumes of spray solution used will be much higher than with stump treatments, usually between 1000 and 3000 litres being required per hectare of rhododendron (excluding gaps between bushes) depending on the size of the regrowth.

6.3.3. Regrowth taller than 1·5 metres usually requires so much spray and labour that its treatment is economically unattractive.

6.3.4. For both stump and regrowth treatment, applications are usually best made using a knapsack sprayer at low pressure, fitted with a cone jet.

6.3.5. Although treated areas may be planted within a few days of application, especially if the crop is not expected to "flush" for about 4 weeks, it is better to leave an interval of one month between application and subsequent planting.

6.4. **AMMONIUM SULPHAMATE SOLUTION, with a concentration of 0·4 kg ammonium sulphamate per litre of water, to all accessible surfaces at any time of the year.**

6.4.1. For cut stumps it is important to saturate all cut surfaces and bark down to soil level. Control can also be improved by treating the soil within 0·3 to 0·6 metres radius of the stump. Applications are best made within 24 hours of cutting.

Plate 1. Atrazine (wettable powder) being applied from a knapsack sprayer.

Plate 2. Paraquat being applied from a knapsack sprayer, using a "Politec" guard to shield the trees.

Plate 3. Applying 2,4,5-T in paraffin to the base of an unwanted broadleaved tree ("Basal Bark" application).

Plate 4. Applying 2,4,5-T to incisions in the base of an unwanted broadleaved tree using a "Jim-Gem" tree injector.

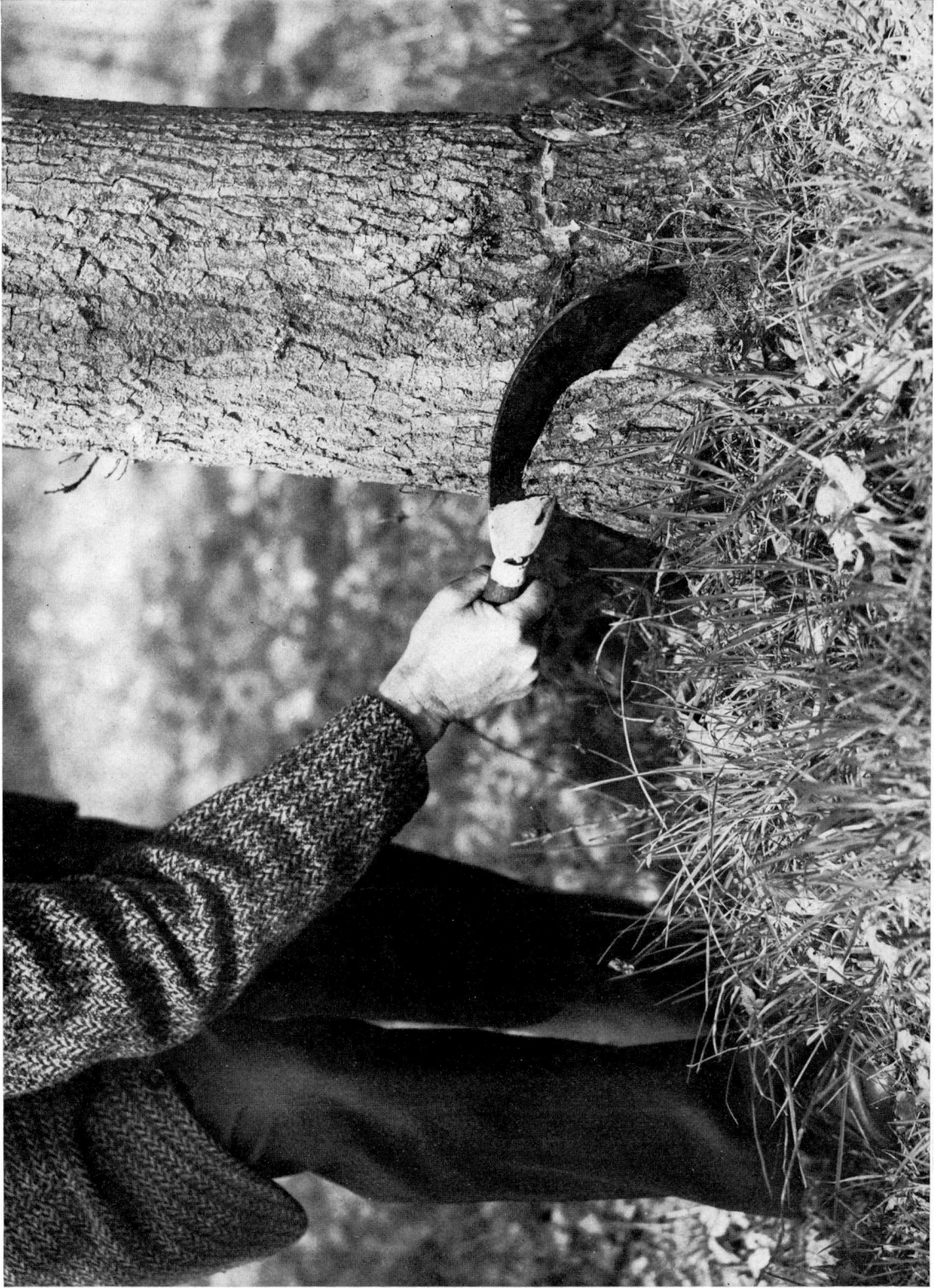

Plate 5. Frill girdling a broadleaved tree prior to application of 2,4,5-T or ammonium sulphamate solution.

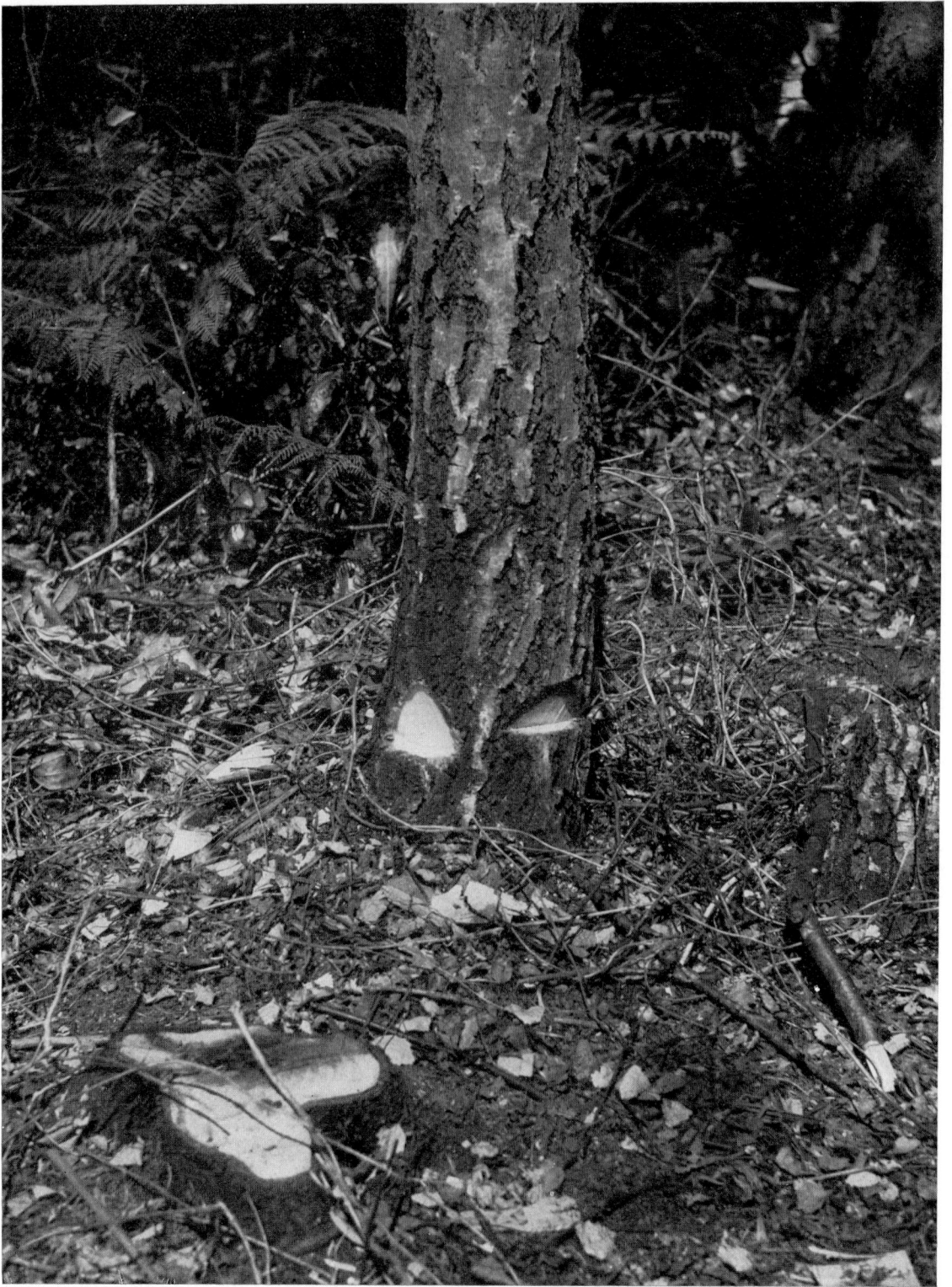

Plate 6. Notches cut in a broadleaved tree prior to the application of ammonium sulphamate crystals.

Plate 7. The effect of 2,4,5-T on freshly cut Sweet chestnut stumps;
(*above*) sprayed with 2,4,5-T.
(*below*) not sprayed with 2,4,5-T.

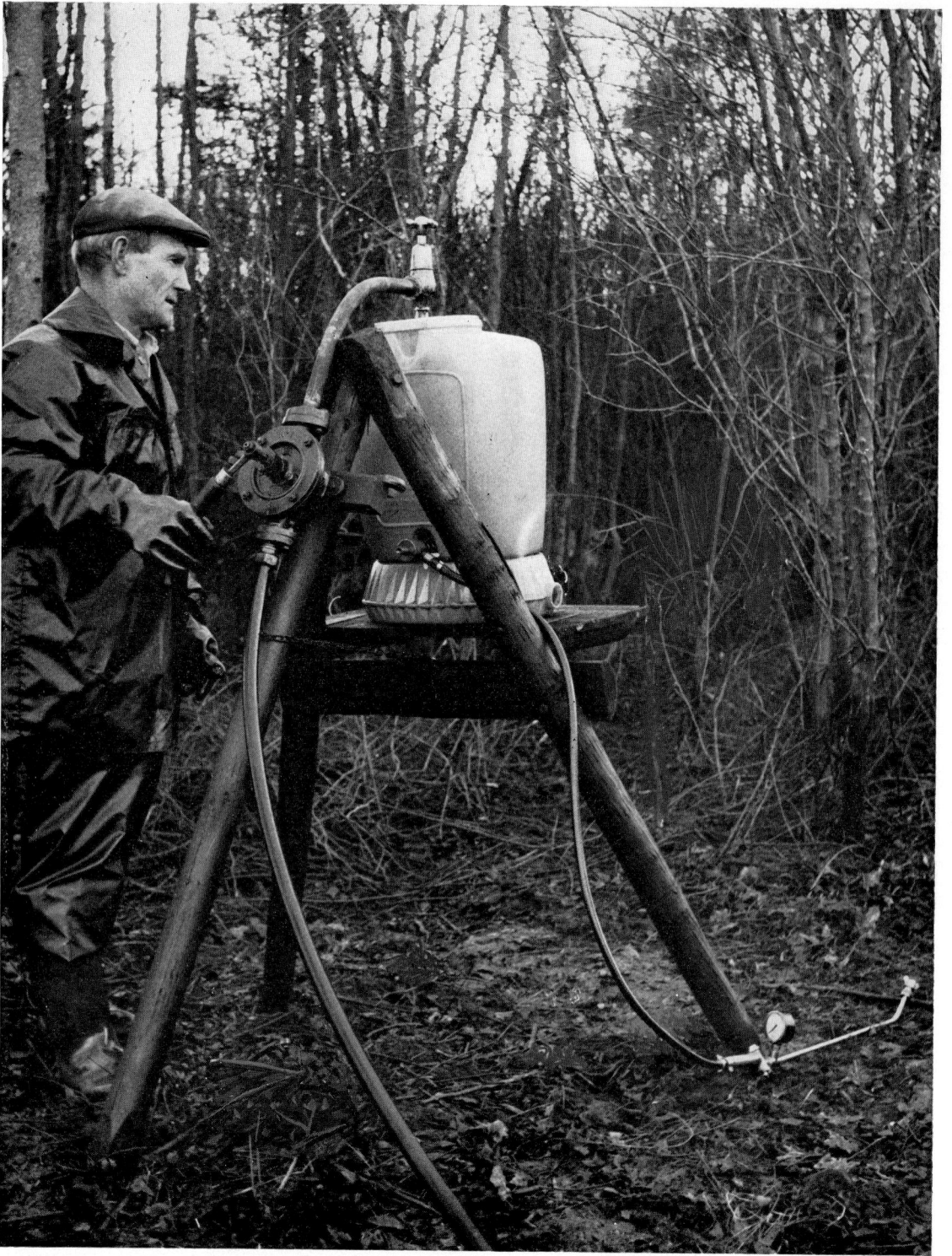

Plate 8. Filling a knapsack sprayer from a semi-rotary pump.

6.4.2.   Coppice shoots and seedling rhododendron can also be effectively and economically controlled, provided they are less than 1·5 metres average height, by an application of this ammonium sulphamate solution to all aerial parts as recommended with 2,4,5–T (see para 6.3.2). For this purpose a non-ionic wetter (see para 12.40) should be added to the spray solution at a rate of 6 mls per litre of spray solution.

6.4.3.   Applications to cut stumps can be made with a plastic watering can or knapsack sprayer fitted with a cone jet. Applications to the foliage, stems and stumps of regrowth are most conveniently made with a knapsack sprayer using a cone jet. Ammonium sulphamate solutions cause severe corrosion of most metals, although corrosion of brass can be reduced by 90% or more by adding 1·0 g of sodium benzoate per litre of spray solution. Corrosion of copper and mild steel by ammonium sulphamate solutions is only marginally reduced by the addition of sodium benzoate. Ideally, therefore, sprayers used for applying ammonium sulphamate solutions should be so designed that the spray solution does not come into contact with the metal parts of the sprayer.

6.4.4.   Areas treated with ammonium sulphamate should not be planted for at least 12 weeks after application as the trees may be damaged by soil residues.

6.4.5.   Post-planting applications of ammonium sulphamate are not recommended.

STUMP AND LOW REGROWTH CONTROL AFTER PLANTING

6.5.   2,4,5–T ESTER in water or paraffin at a concentration of 4·0 to 5·0 litres of 50% emulsifiable 2,4,5–T per 100 litres of diluent (2·0 to 2·5 kg acid/100 litres of diluent) to all accessible surfaces, during the crop's dormant season.

6.5.1.   For stump treatment, applications should saturate to point of run-off all cut surfaces and bark down to soil level, and should be made as soon after cutting as possible, and at least within one week of cutting. A more reliable kill is obtained when paraffin is used as a diluent, but this increases the risk of crop damage.

6.5.2.   Coppice shoots and seedling regrowth can also be effectively and economically controlled with this spray solution using the application method described for pre-planting applications (see para 6.3.2). Since adequate control can be obtained using water as a diluent, this should be used in preference to paraffin.

6.5.3.   For stump, coppice shoots and seedling regrowth, sprays must be placed to miss crop trees, which may be badly damaged by direct contact with the spray solution.

6.5.4.   Applications during the crop's growing season are not recommended because of the high risk of crop damage from such concentrated solutions of 2,4,5–T applied at such high volumes per hectare. Also, hot weather shortly after application can cause volatilisation of the herbicide from treated surfaces, and the vapour can drift onto crop trees. It is therefore important to ensure that the crop is dormant and to stop spraying about one month before flushing is due.

CHAPTER 7

RECOMMENDATIONS VI: SPECIAL WEED SITUATIONS IN THE FOREST

7.1.   The following special weed control problems justify separate, brief sections to draw attention to the main modifications or alterations to the recommendations already given.

CHRISTMAS TREE PLANTATIONS

7.2.   Weed control in Christmas tree plantations needs to be of a very high standard. Loss of needles or deformation due to mechanical interference by weeds, or due to herbicides may be acceptable in normal forest plantations, but are not acceptable for Christmas trees.

7.3.   Norway spruce (*Picea abies*) is the usual tree used for Christmas trees in Britain. This must be taken into account when selecting the herbicide to use.

7.4.   Christmas tree plantations are often planted at close spacing (at about 1×1 metres) and it is more difficult to use herbicides which must be kept off the tree foliage. The use of tree guards on the lance may be unsatisfactory because the spray may hit the trees next to the one being protected.

7.5.   Christmas tree plantations of Norway spruce have been known to respond to the efficient removal of weed competition using herbicides by growing faster. The resultant long internodes at the top of each tree may make them unsuitable as Christmas trees.

Perennial Grasses and Grasses mixed with Herbaceous Broadleaved Weeds

7.6.   Paraquat and dalapon may be used as recommended for forest plantations (see Chapter 2) but see cautionary note in 7·4 above. Atrazine, chlorthiamid and dichlobenil should only be used at the lowest doses recommended in the text (see 2.17, 2.18 and 2.20). These latter three herbicides are all capable of discolouring or browning tree foliage, particularly Norway spruce.

Woody Broadleaves and Mixtures of Woody and Herbaceous Broadleaves

7.7.   Overall applications of 2,4,5 and 2,4–D, as directed for normal forest plantations, can cause browning and loss of conifer foliage. Therefore, these herbicides should only be used at the lowest doses recommended in the text; spraying should be delayed as long as possible after the formation of the terminal resting bud in early August. Avoid spraying in the year of harvesting.

POPLAR PLANTATIONS

7.8.   The growth of poplars can be seriously reduced by weed competition during the first two to three years after planting. Therefore, whilst the trees may appear to be growing quickly and to be be well above the weeds, vegetation for about 1·0 metre radius around the base of each tree should be well controlled.

7.9.   Thick mulches of cut vegetation provide good control and appear to enable the trees to make maximum growth. However, mulches are often too time-consuming and expensive to prepare, and control with herbicides provides a satisfactory alternative.

Perennial Grasses and Grasses mixed with Herbaceous Broadleaves

7.10.   Paraquat and dalapon may be used as recommended for normal forest plantations (see Chapter 2). Chlorthiamid and dichlobenil should not be used as there is some evidence that they depress the growth of poplars. There is no evidence on the use of atrazine.

7.11.   On sites which start free of weeds, simazine at 4·0 kg of a 50% w/w wettable powder per ha (2 kg a.i./ha) at medium volume is an excellent way of preventing weed invasion. However, this is only sufficiently effective on soils with low organic content (say less than 7%). If light weed growth is present at the time of spraying, paraquat can be added to the spray solution and both herbicides applied together.

Woody Broadleaves and mixtures of Woody and Herbaceous Broadleaves

7.12.   Placed applications of 2,4,5–T or 2,4,5–T/2,4–D mixtures can be made around the base of poplar trees, provided direct contact with the bark or foliage can be avoided. The doses used should be as recommended in the text (see Chapter 3).

7.13.   2,4,5–T and 2,4–D formulations can volatilise from treated surfaces in hot weather, and the period of most rapid crop growth (May to July) should be avoided.

FIRE RIDES OR BREAKS

7.14.   Most fire rides or breaks are continuously cultivated or mown. However, where these methods cannot be used, usually because of terrain difficulties, herbicides may be used to control vegetation, either directly or through controlled burning.

Direct Control of Vegetation on Fire Breaks (see Holmes and Fourt, 1961)

7.15.   This method should only be used where vegetation is short or sparse, otherwise death of thick grass swards or scrub may provide a greater hazard than the live vegetation itself.

7.16.   Paraquat (for grasses and herbaceous broadleaves) 2,4–D (heather) and 2,4,5–T (other woody weeds—e.g. gorse) should be used as recommended in the foregoing sections.

Control of Vegetation by Controlled Burning (see Connell and Cousins, 1969)

7.17.   This method is suitable where fairly thick vegetation has become established on a fire ride or break, and is regularly used to control *Molinia caerulea* in Britain.

7.18.   **PARAQUAT at 5 to 10 litres of 20% w/v concentrate per hectare (1·0 to 2·0 kg a.i./ha) in water at medium volume can be applied in July or August to desiccate grasses and other vegetation in fire breaks in preparation for burning off the aerial parts two or three weeks later.**
     The surrounding, undesiccated, green vegetation serves to prevent the fire from spreading and getting out of control. To reduce the risk of the fire spreading beyond the sprayed ride, a strip of surrounding vegetation is often sprayed with sodium alginate solution.

CHAPTER 8

EQUIPMENT

8.1. This chapter summarises the main principles and recommendations for equipment. More details can be found in Forestry Commission Bulletin No 48 (Wittering, 1974).

8.2. The use of the best available equipment and work methods for any particular situation can reduce the costs of applying herbicides substantially, and foresters with large spraying programmes should pay particular attention to these aspects. The choice of the most suitable equipment has to take into account the scale and duration of work, and its accessibility for machines.

DISTRIBUTION OF MATERIALS

8.3. The method of transporting herbicide and diluent on to the area to be sprayed and preparing the spray liquid can affect working costs, especially when MV applications are being made. Where and how to mix the spray solution and distribute it on the site depends very much on local conditions, but everywhere the most economical method is likely to be a compromise between low-cost bulk storage and movement of liquids (or solids for granular herbicides) and the need to reduce "non-spraying" time on the site (walking and refilling). This compromise is usually best achieved by having small portable containers of spray solution well and strategically distributed over the site, which are supplied from a central bulk storage system.

8.4. In most cases the spray solution should be prepared in bulk in a central depot. Here, correct calculation and measuring of quantities is easier and spillage of herbicide concentrates less likely. Moreover, should accidents occur with the concentrate (the main danger in herbicide operations), cleaning facilities and telephones are near to hand.

8.5. It is usually cheaper to mix the spray liquid in bulk in a central depot and then to transport it to the forest site in bulk, rather than to mix small quantities on the site. For many forest programmes, a portable tank holding 500 to 1000 litres prepared in the morning at the depot will provide enough spray liquid to keep the spraying team going for the whole day. For very large programmes employing several spray teams simultaneously, it may be necessary to have larger static tanks in the depot from which the portable tank can be filled. Large static tanks in depots have been widely used for storing paraffin for basal bark or stump spraying because they allow bulk buying of the paraffin at a reduced price.

8.6. Most herbicides will remain dispersed or in solution in the diluent for long periods after mixing, but this point must always be checked before the forest becomes committed to a bulk storage system. For instance, wettable powders like atrazine tend to settle out after mixing, and are not suitable for bulk storage or transport systems unless continuous agitation of the mixture can be arranged. Many commercial tractor-mounted spray tanks for boom sprayers are fitted with recycling systems driven by the tractor P.T.O. system, and thus adequate agitation is provided.

8.7. In the forest, spray liquid can be distributed to various strategic points on roads, rides and within the area to be sprayed using containers of 200 to 500 litres capacity (e.g. 45 gallon drums—205 litres). Storage of spray liquids in flimsy polythene containers in the forest is not recommended because of their vulnerability, although such containers are suitable for storing water (e.g. plastic barrel liners).

FILLING SPRAYERS IN THE FOREST

8.8.   It is usually too costly and often difficult to arrange for the ultimate supply of spray liquid or water to be mounted high enough to allow the sprayer to be filled by gravity. Excellent, small and light, manually operated pumps are available which can quickly pump liquid from the supply (e.g. 45 gallon drum) into the sprayer.

8.9.   Most sprayers used in forestry are carried on the operator's back (knapsack type sprayers) and hold between 9 and 18 litres of liquid. Thus, a full sprayer will usually weigh between 14 and 23 kg and lifting this from the ground on to one's back unaided can be extremely difficult. Portable stands on which to place the sprayer during filling are, therefore, highly desirable. In the Forestry Commission the pump and stand are often combined to make a single unit which the operator transfers from one supply source to another during the day (see Plate 8).

SPRAYERS FOR MEDIUM VOLUME (MV) APPLICATIONS (200 TO 700 LITRES/ha)

8.10.   Sprays at medium volume are commonly used for making foliar applications of herbicides to short grass and herbaceous broadleaved weeds, and for making applications to the stems or stumps of unwanted trees.

8.11.   The sprayers used for medium volume applications produce quite large droplets (100–500 μ diameter) which fall quickly and are little affected by wind. Therefore, accurate placement of the herbicide is possible, facilitating spot or strip spraying around trees or applications to individual stems or stumps of trees. The use of guards to protect crop trees from direct spraying allows the use of herbicides which would otherwise damage the crop (e.g. paraquat, dalapon and glyphosate).

8.12.   The main disadvantage of medium volume applications is the need for large quantities of diluent on the site. The operator spends a great deal of time refilling his sprayer, rather than spraying, and the logistics of getting diluent supplies on the site have to be well thought out to keep "off-spraying" costs to a minimum.

Medium Volume Knapsack Sprayers

8.13.   There are three types of manually operated sprayers:

(1) Continuously pumped knapsack sprayers—which have a small sub-container within or outside the main liquid container which can be easily and quickly pressurised as the operator walks. The sub-container is fed with spray solution continually from the main container. The weight of the sprayer is reduced by using light-weight plastics for the main container, and this saving used to enable more spray solution to be carried (up to 18 litres). This is the recommended type of medium volume sprayer for use in forestry.

(2) Gravity fed knapsack sprayers—the container is made of lightweight plastic and may hold up to 18 litres of spray solution. However, for most forestry operations some pressure is required to operate the jets correctly, and the use of such sprayers is usually restricted to jobs where a dribble-bar gives satisfactory distribution.

(3) Compression knapsack sprayers—in which the whole liquid container is pressurised by pumping air into it before spraying. The container has to be strongly built to withstand the pressure and can only be about two-thirds filled with spray solution because otherwise the pressure would then fall too quickly during spraying. The container should be periodically tested to confirm that it will withstand the pressure. These sprayers are heavy for the amount of spray liquid held, and are not the most economic to use for forest scale operations.

Accessories for Medium Volume Knapsack Sprayers

8.14.   For efficient spraying the following accessories are essential:

(1) *Adjustable pressure control valve.* This should be fitted directly to the outlet from the container to ensure that the spray solution does not leave the container at a pressure higher than the setting. The liquid inside the container must always be at a pressure about 0·25 kg/cm² above the valve setting to ensure a satisfactory flow rate. A visibly reducing or fluctuating flow during spraying indicates that the pressure inside the container is too low.

(2) *Pressure gauge.* This should not be kept permanently on the sprayer, but should be fitted into the lance between the trigger control and the nozzle to check that the jets are discharging at the recommended pressure. With the sprayer filled and held at the normal height, and the lance held as for spraying, the pressure control valve should be adjusted until the pressure gauge reads the correct value.

(3) *On–off trigger on the lance.* This is essential for precise control of the spray.

(4) *Lance.* Normally a short (40–50 cm long) lance is recommended as this allows the operator to hold it comfortably outstretched, with the jet at approximately the correct height above the ground. Longer lances are available.

(5) *Spray jets.* It is essential to fit a jet which gives the spray pattern required, which suits the pressure produced by the sprayer and which emits spray at a rate which is compatible with the feasible walking speed of the operator (usually 2 to 4 km per hour in forest conditions). Suitable types include:

    (a) *Jets producing fan-shaped sprays.* The only suitable ones for forest use are "Floodjets", which operate at 0·5 to 1·5 kg/cm² (7 to 21 lb/in²) and therefore suit sprayers with pressurised sub-containers (see above) which do not easily produce pressures much higher than this. At these low pressures very few fine droplets liable to drift are produced, and "Floodjets" are well suited to making placed foliar applications (spot or strip) in young plantations.

    (b) *Jets producing cone-shaped sprays.* Again, for most forest spraying operations, these should operate at low pressures. For stem or stump treatments with 2,4,5–T, jets capable of giving a fast delivery of a very narrow "pencil" wide stream are generally best, although for coppice stools and big stumps a wider cone can be useful. Adjustable "cone" jets capable of delivering a cone-shaped spray of any chosen width between the two extremes are sometimes fitted. Hollow cone jets are usually used in conjuction with tree guards.

Medium Volume Sledge-Mounted Live-Reel Sprayers

8.15. A full knapsack sprayer can weight up to 25 kg, which is itself a serious disadvantage. A pump, hose and live reel mounted on a sledge ("PHAROS") has been developed by the Forestry Commission to remove this weight from the operator's back.

8.16.   The equipment can be moved by two men to new locations in the forest. Spray liquid is pumped at a controlled pressure by a two stroke engine from a portable storage tank (e.g. a 45 gallon drum) through 150 metres of main hose, which then divides into two 40 metre side lines, each of which terminates in a spray lance and jet. The main line is pulled out from the live reel as the operators work away from the pump into the plantation.

8.17.   This equipment has been found to be more economic than knapsack spraying on difficult terrain or where operators have to walk long distances from the source of the spray liquid.

Medium Volume Vehicle-Mounted Equipment

8.18.   Tractor, Land Rover or other vehicle-mounted spray equipment has so far found little place in forest weeding because of the irregular nature of the topography and obstacles such as stumps and brash etc. Tractors fitted with agricultural spray-booms have occasionally been used to spray grass before planting, and in some areas four hoses and lances have been run directly from a tractor-mounted sprayer to apply paraquat or 2,4,5–T (to stumps) in young plantations.

Guards to Protect Crop Trees during Medium Volume Applications

8.19.   Crop trees must be protected from the sprays of certain herbicides (e.g. paraquat and dalapon), and various guards have been developed for fixing to the lance just above the spray jet to enable these herbicides to be applied to a patch round each tree without directly spraying the tree. These guards are of two main types:

(*a*) *Guards designed to cover the tree.* These are either U-shaped or cone-shaped, nozzles being especially fitted to ensure that when these guards are put up to or over the tree, the guard prevents spray reaching the tree.

(*b*) *Guards designed to cover the spray jet and spray.* These are usually bowl or hood shaped, and are fixed in an inverted position just above the jet so that this lies centrally inside the bowl or hood. When spraying, the inverted bowl or hood is moved round the tree with its lip close to the ground, and thus spray is prevented from reaching the tree.

SPRAYERS FOR LOW VOLUME (LV) APPLICATIONS (90 to 200 litres/ha)

8.20.   Low volume applications are only used in forestry to make overall foliar applications of herbicides to weeds dispersed over the whole area, and for which control in patches or strips around each tree is inadequate (e.g. woody broadleaves, heather, bracken).

8.21.   The main advantage of such sprays is the small volume of diluent required to distribute the herbicide, although this advantage is only exploited to its maximum in Ultra Low Volume spraying (see paras 8.23–8.26). However, the equipment used to make and distribute the spray (mistblowers) produces droplets over a very large diameter range (5 to 800 μ, but most are 100 to 120 μ). The distribution of the majority of the droplets is fairly well controlled, but the smaller droplets can drift for considerable distances and their tendency to swirl and hang in the air necessitates the wearing of complete protective clothing, including a respirator.

Low Volume Knapsack Mistblowers

8.22.   Mistblowers are the only effective type of sprayer currently used for making low-volume applications. Small volumes of spray liquid are delivered from a container into a tube carrying a very fast air stream created by a small engine and fan. This liquid is broken up into very small droplets by the air stream and baffles in the tube, and discharged through a nozzle designed so that the operator can direct the spray. The spray is blown out for distances of up to 5 metres or more, and the operator can achieve fairly uniform coverage of weed foliage up to about 1·5 metres in height for a distance of about 3·0 metres either side of his path, provided he swings the nozzle horizontally in a carefully prescribed way.

Low Volume Tractor-Mounted Mistblowers

8.23.   Mistblowers, operating on similar principles, but mounted on tractors, have been used on a small scale in some lowland forests in Britain. Their use is restricted by bad ground conditions.

SPRAYERS FOR ULTRA LOW VOLUME (ULV) APPLICATIONS (5 to 20 litres/ha)

8.24.   Recently sprayers have been designed which enable very small volumes of liquid to be dispersed satisfactorily, and for those herbicide treatments where the method produces acceptable results, ULV spraying largely avoids the problems of diluent distribution encountered with medium or low volume spraying.

8.25.   ULV spraying machines produce very small droplets of fairly uniform size (mostly within 50 to 150 μ diameter). The uniformity of the droplet size ensures that they all fall under gravity at similar speeds, and the operator can predict where the droplets will settle, given reasonably constant wind conditions. In most spraying operations, a special low volatile oil is used as a diluent to prevent excessive evaporation reducing the size of the droplet between the sprayer and its target, as this would create drift problems.

Equipment for Ultra Low Volume Spraying

8.26.   The type of sprayer which has been used during Forestry Commission trials is shown in the right-hand photograph on the front cover. It consists of a small, battery-powered, spinning disc onto which the spray liquid drops under gravity from a small plastic bottle. The disc is designed so that it throws the liquid off in a fine spray of fairly uniform droplet size, the speed of the disc and the rate of feed of the spray liquid controlling the droplet size. The spray is then allowed to drift and settle under the prevailing wind conditions. The quantity of spray liquid per unit area depends on the walking speed of the operator, the distance between his traverses across the area, and the type of nozzle fitted to the sprayer. Nozzles delivering a nominal 0·5 (yellow), 1·0 (red), 2·0 (grey) and 3·0 (green) mls per second are available for this sprayer.

8.27.   Careful selection of the correct weather conditions and careful training of the operator is required if the best is to be got out of this technique, and if problems of operator contamination and excessive drift are to be avoided. Spraying should not take place in gusty wind conditions, or if constant winds of over 10 m.p.h. persist. A fuller description of the equipment and its operation is to be found in *Forestry and Home Grown Timber*, Volume II(VI) (Brown, Rogers and Thomson, 1973).

DISTRIBUTORS FOR GRANULAR HERBICIDES

8.28.   At first in British forestry, granular herbicides were applied by hand, often with the aid of small "sugar shakers". Such techniques were too laborious, inaccurate and expensive to encourage the wide use of granular herbicides, and mechanised methods were soon developed.

8.29.   Although there are now many machines capable of distributing granular materials, only those using air-flow systems have been used on any scale in forestry. Recently (1973) a gravity-fed machine has been shown to give satisfactory distribution of dichlobenil. Most other types of machine require too much awkward physical effort from the operator during distribution (e.g. turning of handles) to be acceptable on rough forest terrain.

8.30.   Air-flow machines work on a similar principle to mistblowers for liquid sprays. Granules are gravity fed into an air stream created by a small engine and fan, and the air stream carries the granules along a hose and through a nozzle shaped to direct the granules evenly over the area to be treated. The pattern of distribution is often excellent, especially for treating strips of 1 to 2 metres wide as the operator walks. On–off switches in the gravity feed pipe can be used to interrupt the flow of granules between widely spaced trees to save herbicide.

8.31.   Initially, air-flow machines were developed for distributing granules only. More recently machines capable of both mistblowing liquids and also distributing granular herbicides have been developed. Clearly, the greater flexibility of such machines makes them an attractive proposition.

TREE INJECTORS

8.32.   Special tools have been developed for making injections of concentrated herbicide solutions into the vascular tissues of trees. Successful ones are usually of two main types:

(*a*) For injecting at breast height or waist height. The incision is made with a small hand axe using one hand and injections are made with a separate device like a water pistol, which is attached to a small plastic reservoir, with the other hand. A more expensive and intricate tool can be used which ejects the herbicide from the head of the axe as it strikes the tree and makes the incision.

(*b*) For injection at the base of the tree. These usually consist of long, metal tubular shaped reservoirs with a chisel bit at one end. The tube is used as a shaft to drive the bit into the tree just above ground level, herbicide being ejected over the bit and into the incision either automatically as the injector hits the tree or by the operation of a separate lever.

8.33.   Type (*a*) injectors can be used in more confined situations and are less tiring than type (*b*) injectors. However, occasionally trees injected at breast or waist height sprout from the base.

MAINTENANCE AND CLEANING OF SPRAY EQUIPMENT

8.34.   Careful maintenance and cleaning of equipment is the best way of ensuring trouble free operation. Manufacturers' booklets/leaflets and advice from other sources (e.g. Work Study Branch, Forestry Commission) should be read and filed for reference. The following notes give some general points:

(1) Hoses and spray lines should be made from oil-resistant materials, even if the diluent is water, as many herbicide concentrates have oily components.

(2) Spray jets are very easily worn or damaged, and this may have a serious effect on the spray pattern. Jets should be inspected and tested regularly, and replaced if necessary. (After a new jet has been fitted in a sprayer, the rate of delivery of spray solution should be recalibrated).

(3) All equipment must be cleaned thoroughly, **immediately after use.** This is particularly important after using 2,4–D or 2,4,5–T. With these materials, spraying equipment should be cleaned immediately after use by pouring a quantity of paraffin into the tank and recirculating it through the sprayer, pump and spray lines, and back into the tank. This paraffin should then be sprayed out onto waste ground and replaced by a large volume of water containing a wetter or detergent. This should again be circulated thoroughly through the sprayer, and sprayed out on to waste ground. Finally, all parts should be washed through with clean water.

(4) If equipment used for applying phenoxy-type herbicides (e.g. 2,4,5–T, 2,4–D) is not washed out meticulously after use, subsequent sprays may be contaminated and may cause substantial damage. Failure to clean equipment properly is known to have resulted in crop damage and financial loss on farms, forests and in forest nurseries.

(5) Equipment which has been used for ammonium sulphamate solution must be washed and sprayed out thoroughly with clean water immediately after use, and metal parts thoroughly smeared with light engine oil.

(6) **ALWAYS SPRAY OR POUR WATER OR OIL USED FOR WASHING OUT SPRAY EQUIPMENT ONTO WASTE DRY GROUND WHERE IT CAN PERCOLATE INTO THE SOIL. NEVER WASH OUT NEAR WELLS. NEVER POUR WASHINGS INTO DRAINS, DITCHES OR STREAMS!**

CHAPTER 9

AERIAL SPRAYING

General

9.1.   Aerial spraying in forest areas is unique in being the only operation in weed control which is always carried out by contractors who normally have little contact with forestry. Because of this, aerial spraying requires particular care in planning and special organisation. There is widespread public sensitivity about the misuse of herbicides applied from the air, and particular attention should be paid to the safety aspects of spraying—especially to ensure that there is no unwanted drift of spray from the site.

9.2.   Aircraft work quickly but are expensive. Aerial spraying contractors usually have a full programme and have to work one job in with another. The weather is usually the decisive factor, in particular, wind speed and direction. Supervisors and ground markers must be prepared to turn out early or late and must expect last-minute alterations of plans and in particular, delays.

9.3.   At the present time, there is no evidence that helicopters have any technical advantage over fixed-wing aircraft. Which is chosen will depend on the distance of landing facilities from the area to be sprayed and the consequent cost per hectare. In flat country and where convenient landing strips are near at hand, fixed-wing aircraft are likely to be cheaper than helicopters.

9.4.   Of the herbicides recommended in this booklet, only 2,4–D and 2,4,5–T have been widely applied to forest areas from the air, although the technique may well be suitable for applying asulam to bracken. Recently, little or no 2,4,5–T aerial applications have been made because of the increased importance of broadleaves in British forestry and the small size of remaining areas requiring treatment, which are thus more easily sprayed from the ground. Aerial applications of 2,4–D have never been made on a large scale.

Selection of Area and Preliminaries

9.5.   Any area for spraying should be decided 6–9 months ahead of the spray date so that preliminary enquiries can be made and requests to tender or firm negotiations started at least 3–4 months before the date of spraying. Terms should have been agreed 2 months ahead of the operation.

Avoidance of Spray Drift Damage

9.6.   The volume of spray solution used per hectare is low compared with most other spray techniques, and to enable the herbicide to be distributed evenly over all parts of the target droplets have to be small. While the larger droplets will fall to the ground quickly, the smaller ones will drift down slowly. The optimum droplet diameter for spraying is 350μ (0·35 mm) achieved by spray pressures between 1·75 and 2·46 kg/cm² (25 and 35 lb/in²).

9.7.   The aircraft, whether fixed wing or helicopter, has to fly several feet above the highest part of the crop and creates considerable turbulence. There is therefore plenty of opportunity for the wind to pick up the smaller droplets and carry them on to adjoining areas.

9.8.   Winds at ground level are generally least at dawn and at dusk. The first 2 or 3 hours after sunrise and the last hour or two before sunset are times when the risk of drift is least. Wind speeds of up to 15 km/hour (9·3 mph) are normally acceptable. Spraying may also proceed in stronger winds up to 25 km/hour (15·5 mph) if clearly blowing away from susceptible crops.

9.9.   Where blocks of forest near farm land are to be sprayed, details of all crops in a zone up to 1·0 km from the perimeter of the area to be sprayed should be noted, paying particular attention to orchards, seed crops—especially crops of beans or brassicas, glasshouses, Nature Reserves, Sites of Special Scientific Interest, valuable stock and beehives. If susceptible crops or other interest exist the manager should only continue with plans to spray from the air if he is confident that the safety of whatever might be at risk can be assured. A map on the scale 1:10,000 (6 inches to the mile) is ideal for recording these details. As part of the pre-spray routine, most contractors carry out a survey from the air or by Land Rover. However, it is in everyone's interest for local forest staff to supplement the contractor's survey information from local knowledge.

9.10.   It may be desirable to leave a boundary strip, for example, to conceal the brown foliage of treated weeds in an area of high amenity, or to leave a safety margin between the sprayed area and susceptible crops. At least 300 metres should separate susceptible crops or stock from the nearest area to be aerially sprayed, and this distance may have to be greater if the wind direction is unfavourable.

9.11.   The Nature Conservancy Regional Officer for the locality must be informed at the very first opportunity if land including or adjacent to nature reserves or Sites of Special Scientific Interest (S.S.S.Is) are to be sprayed from the air.

The Contract

9.12.   When arranging an aerial spraying contract, the following points should be taken into account:

(1) *Dose and formulation of material to be used*. Often there is only one formulation commercially available. However, where there is any choice, the less volatile formulation should always be selected, see para 3.4.

(2) *Date of spraying*. Limits to the period during which spraying may take place should be clearly agreed. Otherwise contractors are very liable to make applications when it suits them, and this date may be unsuitable for good control or for crop safety.

(3) *Third-party claims*. Normally it is best to ensure that the contractor provides third-party cover, as safe application of the herbicide depends very much on the skill of the pilot. In any case, it must be agreed both who is responsible in the event of damage to neighbouring crops and the procedure to follow should a claim be made. All staff must clearly understand that nothing other than a bare acknowledgement of receipt of a claim should be made until the claim has been fully investigated.

(4) *Responsibility for spraying*. Normally this responsibility will be carried by the pilot because he will normally have much more experience of aerial spraying than any forester. However, the forester or his agent should have the power to prevent spraying when weather conditions are clearly unsuitable. For instance if wind speed is over 25 km/hour (15·5 mph) or if rain is falling or is quite definitely forecast for the next few hours, the power to veto the operation is valuable.

(5) *Services required*, e.g. location of and access to landing strip; supply of diluent; storage tanks or bowsers; refilling procedure; radio contact between landing strip and spray area; ground markers, how many, where and how controlled.

Arrangements between Signing Contract and Spraying

9.13.   One supervisor (and a deputy if possible) should be nominated to be in charge of the ground operation, and his duties and responsibilities should be clearly defined. Because of the likelihood of short-notice changes of plan, this man should be freed from other definite commitments during the period when spraying is likely to take place. He must also be free to take part in any planning discussions beforehand.

9.14.   If ground markers are required, the distance between consecutive flight paths, likely directions of flight and any special arrangement must be discussed with the contractor and a procedure agreed, sufficiently before the event, for the men involved to be instructed and rehearsed.

9.15.   For even spray distribution, the aircraft must fly uniformly-spaced parallel courses over the whole area. Ground markers can be of the greatest assistance, but the men handling the markers must be clearly instructed so that there is an unambiguous indication of the next line of flight immediately one line is finished. Markers must be clearly visible from the air from all angles. Men handling markers must know what to do if an aircraft runs out of spray, part way down a run.

9.16.   It should be agreed with the contractor who will be responsible for informing neighbours and other interested parties such as shooting tenants and beekeepers, and the nature of information given. Normally, the approximate date of spraying and the chemical to be used should suffice.

9.17.   A meeting between the contractor and the forester or other member of the staff who is to be responsible for ground operations should be planned for 10–14 days before the expected date of spraying, to go over the details of the final arrangements. The survey of adjacent crops should have been completed by this date and a list made of neighbours who have been informed that spraying is likely.

CHAPTER 10

COSTS OF WEEDING BY HERBICIDES

10.1. Table 4 gives the relative direct costs (excluding overheads) of the various herbicide treatments. The herbicide and labour costs used were those prevailing in July, 1974; retail costs for relatively small quantities of the herbicide products have been used, and foresters who use large quantities of a particular herbicide should be able to buy more cheaply.

10.2. It is impossible to give absolutely accurate costs for all situations. The table should be taken as giving a reasonable idea of the comparative costs of using the different herbicides/techniques, but should not be relied on to give an accurate costing for any particular operation.

10.3. For comparisons between the costs of using herbicides and using other methods (hand or mechanical cutting tools), see Forestry Commission Bulletin 48 (Wittering, 1974).

Weed Type	Treatment	Herbicide	Rate per treated Hectare	Area Treated
Grass and/or herbaceous broadleaves	Foliar (sprays) or Surface (granules)	Atrazine	8·0 kg product (4·0 kg a.i.) 8·0 kg product (4·0 kg a.i.)	P(1·0 m patches) S(1·0 m wide strips)
		Chlorthiamid	50·0 kg product (3·75 kg a.i.) 50·0 kg product (3·75 kg a.i.)	P S
		Dichlobenil	50·0 kg product (3·75 kg a.i.) 50·0 kg product (3·75 kg a.i.)	P S
		Dalapon	13·5 kg product (10·0 kg a.i.) 13·5 kg product (10·0 kg a.i.)	P S
		Paraquat	5·5 litres product (1·1 kg a.i.) 5·5 litres product (1·1 kg a.i.)	P S
Bracken	Foliar Foliar or surface	Asulam Dicamba	10·0 litres product (4·0 kg a.e.) 10·0 litres product (4·0 kg a.e.)	0 (Overall) 0
Heathers	Foliar	2,4-D 50% 2,4-D 40% special ULV formulation	8·0 litres product (4·0 kg a.e.) 10·0 litres product (4·0 kg a.e.)	0 0
Woody broadleaves	Foliar	2,4,5-T 50% 2,4,5-T 30% special ULV formulation	6·0 litres product (3·0 kg a.e.) 7·0 litres product (2·1 kg a.e.)	0 0
	Stem/stump	2,4,5-T 78%	200–600 litres/ha of a solution containing 1·5 kg acid per 100 litres of paraffin	Bark at base of stem ("basal bark")
			200–600 litres/ha of a solution containing 1·5 kg acid per 100 litres of paraffin	Frills + "basal bark"
			200–600 litres/ha of a solution containing 1·5 kg acid per 100 litres of paraffin	Cut surface and surrounding bark of stumps.
		2,4,5-T 50%	Assume 6 injections per tree of 1 ml–500 trees/ha	Injection into stems.
		2,4-D 50%	Assume 6 injections per tree of 1ml–500 trees/ha	Injection into stems.
		AMS	200–600 litres/ha of solution containing 0·4 kg AMS/litre of water.	Frills + basal bark cut surface of stumps.
Rhodo-dendron	Foliar, stem and stool	2,4,5-T	300–1000 litres/ha of a solution containing 2·0 kg acid per 100 litre of water.	Whole bush.
	Stump	2,4,5-T	200–600 litres/ha of solution containing 2·0 kg per 100 litres of paraffin.	Cut surface of stump and surrounding bark.
		AMS	200–600 litres/ha of a solution containing 0·4 kg AMS per litre of water.	Cut surface of stump and surrounding soil.

Notes: 1. Herbicide costs: 1974 prices for small quantities of products.
2. Labour costs: Assume piece-work rates based on £25 per week basic wage + 30% incentive. (40 hour week).

Herbicide (assuming 2·1 × 2·1 m spacing)	Costs (£/ha) Application costs (Labour + Direct oncost + machinery costs where applicable)				
	med. vol.-tractor	med. vol.-knapsack	low vol.-mistblower	ULV	other application methods
2·40	—	3·40	—	—	
5·10	—	4·50	—	—	
7·60	—	—	—	—	2·40–3·00 ⎫
15·80	—	—	—	—	2·70–3·40 ⎬ Air flow
10·10	—	—	—	—	2·40–3·00 ⎰ machines
21·10	—	—	—	—	2·70–3·40 ⎭
2·50	—	7·50	—	—	—
5·20	—	—	—	—	—
2·40	—	7·50	—	—	—
4·90	—	—	—	—	—
20·30	—	6·80–13·30	5·00–12·00	—	—
37·40	—	6·80–13·30	5·00–12·00	—	—
14·90	—	8·80–10·10	8·00–10·10	—	—
10·00	—	—	—	7·50–8·75	—
14·80	1·00–2·00	6·80–13·30	5·00–12·00	—	—
10·30	—	—	—	2·80	—
17·00–51·00	—	12·50–22·50	—	—	—
17·00–51·00	—	12·50–22·50	—	—	—
17·00–51·00	—	12·50–22·50	—	—	—
7·40	—	—	—	—	5·60–7·30 ⎫
					special injection equipment
5·60	—	—	—	—	5·60–7·30 ⎭
80·00–240·00	—	12·50–22·50	—	—	—
80·00–240·00	—	12·50–22·50	—	—	—
30·00–100·00	—	15·00–35·00	—	—	14·50–32·00
20·00–60·00	—	12·50–22·50	—	—	12·20–21·60
80·00–240·00	—	12·50–22·50	—	—	—

3. Direct oncosts: Charged at £2·50 per day.
4. Machinery costs: 1974 rates for depreciation plus maintenance.

CHAPTER 11

SAFE USE OF HERBICIDES

11.1.   The chemical industry and Government take considerable care to ensure that each herbicide is safe to use. Therefore, their recommendations should be followed. Forest managers, supervisors and operators must accept responsibility for ensuring that herbicides are safely applied in accordance with the relevant recommendations.

11.2.   Managers particularly, and to a lesser extent supervisors, should know the part played by the chemical industry and the Government, for only in this way will they understand where their responsibilities begin and their extent. The following paragraphs give a summary of the various schemes and legal aspects concerning the testing and use of herbicides, and every manager and supervisor should be familiar with these details.

LEGISLATION, THE LAW AND VOLUNTARY SCHEMES CONCERNED WITH THE SAFE USE OF PESTICIDES

ACTIVITIES OF CHEMICAL INDUSTRY AND GOVERNMENT

Pesticides Safety Precautions Scheme (P.S.P.S.)

11.3.   In Britain firms wishing to market a new pesticide, a new formulation of a pesticide, or wishing to extend the uses of a pesticide are requested to have all aspects of its safety considered under the P.S.P.S., which is administered by the Ministry of Agriculture, Fisheries and Food on behalf of the Government. Although this scheme is voluntary, co-operation between Government and Industry has been excellent since its introduction in 1957.

11.4.   Under the Scheme, the onus is on manufacturers, distributors and importers who wish to introduce a new pesticide or new use for a pesticide, to produce details and results of research to show that the product is safe for its proposed use. The Ministry of Agriculture publishes a booklet which describes the Scheme, the type of tests required, and which indicates how companies can obtain further advice on the type of research to carry out for particular anticipated uses. Of course the type of research carried out is influenced by the type of risk expected. An expert committee considers the results of all these tests and decides whether to give the product clearance for its proposed use. The committee keeps products under review after clearance to see if substantial usage throws up any further problems.

11.5.   From the user's point of view, the main end product of this committee's deliberations is the publication of a *Recommendation Sheet* which sets out the limitations to its use and precautions that should be taken when using the pesticide. The firm marketing the product also undertakes to show these limitations and precautions on the label stuck to the concentrate container. This is why users should always keep the concentrate in its original container.

11.6.   Copies of the booklet describing the scheme and the Recommendation Sheets can be obtained from the Ministry of Agriculture, Fisheries and Food, Pesticides Branch, Great Westminster House, Horseferry Road, London SW1 24E. Managers and supervisors concerned with everyday use of pesticides are recommended to obtain copies of those Recommendation Sheets covering the products they use.

Agricultural (Poisonous Substances) Regulations

11.7.   Legislation dealing with the safe use of agricultural pesticides is made under the Agricultural (Poisonous Substances) Act 1952, as extended by the Agricultural (Poisonous Substances) (Extension) Orders 1960–1966. The object of this Act is to protect employees against the risk of poisoning from the more toxic of the pesticides used in any form of land husbandry.

11.8.   Under this Act, Agricultural (Poisonous Substances) Regulations are issued from time to time— the latest being the Agriculture (Poisonous Substances) Regulations 1966 to 1969. Substances which are subject to restrictions are listed and the details of the restrictions and protective clothing necessary are given. Any employer who does not ensure that these restrictions are followed and that the protective clothing is worn is liable to prosecution.

11.9.   At the present time none of the herbicides recommended in this booklet is considered sufficiently toxic to justify being included in these Regulations.

Pharmacy and Poisons Act, 1933

11.10.   This Act states that recognised poisons should only be sold by authorised or registered sellers. The Act also requires (a) that an official list of poisons should be kept by the Government (the latest list is contained in the *Poisons List Order* 1972 [Statutory Instrument 1972, No. 1938] as amended by the *Poison List Order* 1974 [S.I. 1974, No. 80] and *Poisons List [No. 2] Order* 1974 [S.I. No. 1556] and (5) that conditions of packing, labelling, transport, storage and sale must be specified (the latest specifications are contained in the *Poisons Rules*, 1972 [S.I. 1972 No. 1939] as amended by the *Poisons [Amendment] Rules* 1974 [S.I. 1974 No. 81] and the *Poisons [Amendment] [No. 2] Rules* 1974 [S.I. 1974 No. 1557].)

11.11.   Paraquat is the only herbicide in this booklet covered by this Act.

RESPONSIBILITIES OF USERS OF PESTICIDES

11.12.   The law expects that people who deal with things which are potentially dangerous in themselves will exercise a very high standard of care to ensure that others do not come to harm.

Employer's Responsibility to his Employees

11.13.   An employer owes to his workmen a duty to ensure that a safe system of working is adopted. What constitutes a safe system depends on the circumstances. Many situations are not clear cut, but the judgement could well rest on whether the employer had taken account of any authoritative advice available (e.g. through the Pesticides Safety Precautions Scheme).

Occupier's Responsibility to his Neighbours

11.14.   If an occupier of land uses chemicals to spray his crops and the herbicides harm his neighbour's crops or livestock, he may be liable for any damage caused. If the spraying was done by a contractor the occupier may still be liable. In such a case he may be able to recover damages from the contractor if he can show that the damage occurred through the contractor's negligence, provided such indemnity is not excluded by the contract. An occupier of land is under a duty to fence in his stock but he is usually under no obligation to fence out his neighbour's stock unless the lease or tenancy imposes on him this responsibility. Consequently, in general no legal redress exists if poisoning occurred while the livestock was trespassing on the land which had been sprayed; but the position may be different if a fence was known to be weak, the chemical used was known to be poisonous to livestock and the occupier failed to issue a warning.

Occupier's Responsibility to the Public

11.15.   Both an occupier of land and a contractor owe a duty to members of the public, e.g. persons lawfully passing along the highway (road, footpath or other public right of way) which crosses land being sprayed, to ensure that they are not injured by chemicals which are used. This responsibility also extends to persons visiting the land for the purpose of trade or business and probably also to purely social visitors. This responsibility does not extend to trespassers, although the law makes an exception of young children, from whom it does not expect the same standard of behaviour as adults. If spraying operations constitute an attraction to children, the person carrying out the operation is responsible for ensuring that effective measures are taken to exclude children.

Where herbicides are sprayed onto wild fruit which the public may pick, for example bilberries, blackberries and elderberries, the members of the public should be made aware of the dangers by means of well-sited, easily-read signs.

SAFE USE OF HERBICIDES IN PRACTICE

GENERAL CONSIDERATIONS

11.16.   Most of the herbicides recommended in this booklet are of low mammalian toxicity, and none is known to have chronic long-term side effects on people who are exposed to the herbicides during normal spraying operations.

11.17.   Concern is often expressed about the detrimental effects of herbicides on wildlife. This concern is greatest when overall spraying is to be carried out, for strip and patch spraying leave 50 to 75% of the area untreated. However, where overall spraying has been completed (e.g. 2,4,5–T or 2,4–D on woody weeds) there has been no evidence of ill-effects to wildlife other than to those plants known to be susceptible to the herbicide and to animals directly dependent on these species as a food source, and then only on, or in the immediate vicinity of, the sprayed area.

11.18.   All the herbicides recommended in this booklet are considered to be of short or moderate term persistence in the environment, and none is known to accumulate in any biological system.

11.19.   Some of the herbicides recommended here may find their way into streams and rivers at very low concentrations. The risk only arises shortly after application and if the herbicides are properly used studies have shown that these concentrations are too low and the pollution too ephemeral to cause problems of toxicity to humans or to fish.

11.20.   The main risk of damage to humans, neighbours' crops, stock and domestic animals, or wildlife, arises through accidents or failure to follow good, safe, spraying methods. Human error cannot be eliminated completely—nor is the safety of the herbicides recommended here critically dependent on such perfection—but sensible attention to training and supervision will ensure that accidents are very rare.

11.21.   The following sections make specific recommendations on safe working procedures with this background in mind.

PROTECTION OF THE OPERATOR

Recommendations made under the Pesticides Safety Precautions Scheme

11.22.   The requirements given in the recommendation sheets issued under the Pesticides Safety Precautions Scheme (P.S.P.S.) are given below. These requirements are also printed on all containers in which the herbicides are sold.

Ammonium sulphamate: (Unnumbered sheet issued 8.2.65). Wash concentrate from skin or eyes immediately. Avoid working in spray mist. Wash hands and exposed skin before meals and after work.

Asulam: (Recs/929 issued 1.7.72). Wash hands before meals and after work.

Atrazine: (Recs/897 issued 1.4.72). Wash hands before meals and after work.

Chlorthiamid: (Recs/680 issued 1.8.70). Wash hands before meals and after work.

Dalapon: (Recs/405 issued 1.6.68). Dalapon is irritating to the eyes and can be to the skin. Remove heavily contaminated clothing immediately. Wash splashes from skin or eyes immediately. Wash hands and exposed skin before meals and after work.

Dicamba: (Recs/203 of 1.2.67.). Wash concentrate from skin or eyes immediately. Wash hands and exposed skin before meals and after work.

Dichlobenil: (Recs/695 issued 1.9.70). Wash hands before meals and after work.

2,4–D (Recs/1043 issued 1.12.73). Wash concentrate from skin or eyes immediately. Wash hands and exposed skin before meals and after work.

Paraquat: (Recs/942 issued 1.9.72). Paraquat is subject to the Poisons Rules, but is not in the Agricultural (Poisonous Substances) Regulations. **Wear protective gloves and face-shield** when handling the concentrate. Wash concentrate from skin or eyes immediately. Avoid working in spray mist. Wash hands and exposed skin before meals and after work. Remove heavily contaminated clothing immediately.

2,4,5–T: (Recs/1042 issued 1.12.73.). Wash concentrate from skin or eyes immediately. Wash hands and exposed skin before meals and after work.

Oil diluents: No sheets are issued covering the use of oils as diluents. However, a number of Government leaflets (e.g. "Industrial Dermatites" Form SWH 2064 by Her Majesty's Factory Inspectorate) draw attention to the risk of dermatitis if working with oil, especially diesel oil for long periods. Experience also suggests that some workers suffer respiratory discomfort when working with diesel oil for long periods. To reduce these risks (a) the wearing of protective gloves and the use of barrier cream is recommended when working with 2,4,5–T in oil and (b) premium grade paraffin is now recommended as the oil diluent rather than diesel oil because it is less likely to cause skin and respiratory complaints.

11.23.   Since none of these recommendations comes within the scope of the Agricultural (Poisonous Substances) Regulations, they cannot be enforced by processes of the law. Nevertheless, it is suggested that it should be the unavoidable duty of a good employer to implement these recommendations (see para 11.13 above.)

11.24.   Recommendations under the P.S.P.S. are generalised, and each employer is left to decide what practical measures are necessary to fulfil them, taking account of the actual circumstances in which his employees work. Most of the recommendations can be met by providing washing or cleansing facilities and protective clothing, and by commonsense instructions on working procedures.

Cleaning Facilities and Protective Clothing

11.25.   For all herbicides, the P.S.P.S. recommendations sheets also state that, in all circumstances, workers should wash before meals and after work. Thus soap and water must be available in the forest, or a waterless skin cleanser and paper towels. A source of supply of suitable materials is given in Chapter 12.

11.26.   For most herbicides the P.S.P.S. recommendations sheets lay down that splashes of any concentrate must be washed from the skin or eyes immediately. Generally, it is safer to mix spray solutions in forest depots (see paras 8.3 to 8.7) where washing facilities are available. Accidents with concentrates form the major risk to operators and to the environment, and help is usually more immediately available in a forest depot. Should depot mixing be impractical, goggles or face shield should be worn to protect the eyes when handling the concentrate.

11.27.   Other specific requirements of the P.S.P.S. recommendations are met by wearing protective clothing. Table 5 below lists suitable protective clothing, which should not only fulfil these recommendations, but which should also reduce the risk of operator discomfort. The proposals in the table are based on the following assumptions:

(1) That workers concerned are likely to remain on such work for several days at a time.

(2) That spraying of vegetation at medium volume rates will be done at low pressure with the jet held no higher than knee-height. In such circumstances, the worker will brush past freshly treated vegetation and so pick up spray solution on his legs and feet; hence the need for water-proof (and sometimes oil-proof) trousers and boots if paraquat is sprayed onto grasses not more than about 20 cm tall, water-proof trousers are not essential if Wellington boots are worn.

(3) That low-volume and ultra low-volume sprays will be applied from the ground by mistblower or ULV applicators or by aircraft, and droplets are likely to cover the vegetation at least up to waist height and sometimes higher. In gusty winds, occasional swirls of droplets may envelop the operator (or ground marker in aerial spraying), even though the risk is minimised by working so that the spray is directed down-wind from the position of the operator.

(4) All spraying equipment is maintained in good order with well-fitting, leak-proof filling lids, and joints that do not leak, and that care is taken in filling not to get spray solution outside the container.

(5) For operations such as handling concentrates, protective clothing is essential for the safety or well-being of the operator. However, for other operations it is less essential and is primarily to reduce the risk of discomfort. As some of the protective clothing can itself be uncomfortable, for example on difficult terrain in hot weather, it is important to distinguish between what is essential and must be worn in all conditions, and what a supervisor may allow to be taken off at his discretion. Table 6 is laid out with this distinction in mind.

11.28.   It is recommended that respirators are made available for all low volume or ultra low volume spraying operations, in spite of the fact that the P.S.P.S. recommendations sheets only require protection from spray mist with paraquat. This recommendation is made on the basis that direct contamination of the operator's respiratory system with any herbicide or oil diluent is undesirable, and that there are occasions when this cannot be avoided with these spraying techniques without the use of a respirator. Paraquat should never be applied at low volume or ultra low volume.

11.29.   The P.S.P.S. requirement of a face-shield when handling paraquat concentrate can be met by using goggles such as are listed in para 12.52, provided means are instantly to hand to wipe off splashes from skin of the face and mouth. Otherwise a face-shield should be worn.

11.30.   Strong, plastic gloves are recommended as essential for all spraying operations, and particularly when handling paraquat concentrate (see P.S.P.S. Recommendations). Thin, rubber, unlined gloves are too easily torn to make them a practical proposition when using herbicides in forest conditions. However, it is vital that the lined gloves recommended are examined regularly and frequently to see that they are sound and clean. They achieve the exact opposite of what is intended if the lining absorbs any herbicide and holds it against the operator's skin. Another alternative is thin polythene gloves which are used only once.

11.31. Barrier cream rubbed into the skin and hands and wrists reduces the risk of dermatitis and is recommended when using oil for long periods, especially by operators whose skin is easily inflamed.

Protection of Neighbouring Crops

11.32. The greatest risk of damaging neighbouring crops arises from spray drift (but see also para 11.33). This risk only arises from low volume and ultra low volume applications of herbicides. Asulam, dicamba, 2,4–D and 2,4,5–T may be applied in this way; the widespread use of the latter two herbicides and the great range of crops they could damage means that special care should be taken when using them. Particularly vulnerable (and highly expensive) crops are in orchards, glass-houses and field crops being grown for seed. The main points to observe are:

(1) Avoid spraying in areas immediately adjoining susceptible agricultural or other crops, or leave a safe buffer between them and the sprayed area. This buffer can be sprayed at medium volume with a knapsack if necessary.

Suitable buffers are 50 to 100 metres for mistblowers or ultra low volume applicators, and 400 to 500 metres for aerial applications.

(2) Do not spray on very windy days. Wind speeds should be less than 15 km/hour (9·3 m.p.h.) and preferable less than 10 km/hour (6·2 m.p.h.).

(3) Keep spray "nozzles" as near to the target as consistent with good distribution.

11.33. With ester formulations of 2,4–D and 2,4,5–T considerable risk of damage to neighbouring crops can also arise from volatilisation of the herbicide from treated surfaces, the vapour "cloud" so formed drifting for long distances. Damage has been recorded on crops 800 metres from the site of spraying.

11.34. This type of risk can be considerably reduced by using low-volatile ester formulations (esters equal to or of lower volatility than the iso-octyl ester), and by restricting spraying to periods when the daily maximum temperature is unlikely to reach 20° Centigrade. Risk of damage from volatilisation is greatest with treatments requiring heavy rates of 2,4,5–T or 2,4–D (e.g. Rhododendron or basal bark treatments) and where oil is used as the diluent.

PROTECTION OF WILDLIFE, LIVESTOCK, DOMESTIC ANIMALS AND WILD PLANTS

Animals

11.35. As far as is known none of the herbicides, used *as recommended in this booklet* will have any direct ill-effect on mammals, birds, insects etc. It is, however, important to realise that data on the susceptibility of animals to a pesticide has only been collected on species believed to represent the major groups in the animal kingdom, with particular attention being paid to species considered important to man.

11.36. Some indirect effects are inevitable—such as the destruction of food plants on which animals rely, but many of these plants would have died naturally in the course of normal forest growth; the use of a herbicide will only have hastened an inevitable process. If certain animals are to be protected or encouraged then the most important question is what type of forest is required rather than whether pesticides are to be used.

11.37. Calculations show that the risk of animals ingesting lethal doses of any of the herbicides recommended in this booklet by licking or eating treated material is negligible. The major risk of poisoning animals arises from accidental spillage or contamination of static drinking water (e.g. puddles, ponds). Great care should be taken to avoid accidental spillage; herbicide concentrates should *never* be left in the forest overnight or over the weekend in case they should fall into the wrong hands. Care should always be taken to avoid contamination of natural puddles and water courses, as well as drains and reservoirs (see also para 11.42–11.47), as these often form sources of drinking water for animals, wild and domestic.

11.38. Plants which are poisonous to livestock and which are normally unpalatable, in particular ragwort (*Senecio jacobaa*), may become palatable following foliage spraying. All livestock must be excluded from sprayed areas where such weeds may be growing for 3 to 4 weeks after spraying.

Fish

11.39. Many species of fish are sensitive to pollution of their water with herbicides, although if the constraints and disciplines recommended in this leaflet (see particularly para 11.42–11.47 below) are accepted there should be no risk of dangerous levels of pollution.

11.40. Of the herbicides recommended in this booklet, 2,4–D and 2,4,5–T are the only ones likely to create a risk. These two herbicides may sometimes be applied overall to quite a large area in a single operation, and it is known both that small quantities may find their way into streams and rivers, and that fish vary in their sensitivity to different esters of both herbicides (Turner, *in draft*). Of the commonly available esters, butyl esters are more toxic than most. Thus, besides adhering to the constraints recommended in paras 11.42–11.47 below, it is suggested that butyl esters of 2,4–D or 2,4,5–T should not be used in situations where the risk of stream or river contamination is high.

Plants

11.41. Many plants, other than those it is necessary to control, will be susceptible to the herbicides recommended in this booklet. If rare or interesting plants or plant communities occur in the forest area, then their protection must be taken into account. In those areas affected, the use of certain herbicides may have to be reduced or forbidden if this objective is to be achieved. In Nature Reserves or Sites of Special Scientific Interest, expert advice should always be sought beforehand on the likely effects of any weed control measures.

PROTECTION OF WATER SUPPLIES

11.42. In all spraying operations, the greatest care must always be taken to avoid polluting water-courses or drinking water supplies, either when spraying or when cleaning up afterwards. Particular care must be exercised with overall sprays of 2,4–D, 2,4,5–T or asulam in areas where the water supply of farms or hamlets may come untreated from nearby streams, and in areas close to reservoirs.

11.43. Overall foliage sprays of 2,4–D, 2,4,5–T or asulam provide a greater risk of pollution than the herbicides used for grass/herbaceous broadleaved weed control, which are often used on small areas at a time, and which are usually applied to only a proportion of the area. Forestry Commission trials with 2,4–D and 2,4,5–T confirm that pollution from spraying operations is normally at a very low level, but that levels greater than the 0.01 parts per million mentioned in para 11.47 below can occur in streams draining directly from the sprayed area in the first few days after spraying, particularly if heavy rain falls (Aldhous, 1967; Anon, 1969).

Although the risks of poisoning human beings (or livestock) with drinking water contaminated during the course of any treatment recommended here are negligible, such materials can make the water taste unpleasant. Therefore, on no account should any herbicide be sprayed onto open water surfaces in a ground spraying operation, nor should water used to wash out containers or spray equipment ever be poured into streams or down drains. In aerial spraying operations it is impossible to avoid spraying some open water surfaces; therefore the risk of tainting should be reduced by following the instructions in para 11.47 below.

11.44.   The greatest risk to water supplies of man and beast is through accidental spillage, leaking drums or carelessness when disposing of or washing out containers. Such accidents or malpractices could lead to far higher local concentrations of herbicides than could possibly result from well-conducted spraying operations. It is essential therefore to ensure that accidents and carelessness are avoided. Containers of concentrate or spray solution should never be left in the forest overnight or over the weekend, as this exposes them to the risk of damage from vandals.

11.45.   In ground-water catchment areas (i.e. areas where most of the surplus precipitation may percolate through permeable strata below the rooting depth of plants into wells and boreholes to supply water for domestic or industrial needs), herbicides may be used with little risk, provided attention is paid to the points made in paras 11.42–11.44 above. Studies have shown that all the herbicides recommended here break down in the soil and neither accumulate nor are easily leached from the soil (Audus, 1960; Burschel, 1963). Other studies have shown that the oils used to dilute 2,4,5–T or 2,4–D are unlikely to penetrate more than a few inches into the soil and do not themselves constitute a threat to ground-water supplies (Linden et al., 1963). However, should the area to be sprayed be rented or leased from a water authority, their agreement to the operation must always be obtained beforehand.

11.46.   In surface-water catchment areas (i.e. localities where run-off drains into reservoirs or into streams supplying farms or hamlets with water), there is a greater risk of surface-water run-off carrying herbicide into water supplies, and in addition to following a good safe spraying procedure, it may be necessary to restrict the proportion of the total catchment area sprayed in any one operation (see para 11.47 below). If the area is rented or leased from a water authority, their agreement to the operation must always be obtained beforehand. Many water authorities are unfamiliar with the techniques of applying herbicides, and their natural reluctance is often overcome by taking their water engineer or chemist with a competent spraying specialist to the site, and explaining the method of handling and applying the herbicide.

11.47.   The Water Research Association recommend that the content of 2,4–D, 2,4,5–T and asulam in water supplies at consumption point should not exceed 0·01 p.p.m. Many water authorities are members of the Water Research Association, but it is important to realise that they are not bound to accept this recommendation, and may not want to accept the risk of pollution at any level. If 0·01 p.p.m. is accepted as the maximum allowable concentration of these herbicides, by making various assumptions it is possible to calculate the maximum proportion of a catchment area that should be sprayed to avoid any risk of exceeding this level. For instance, it is useful to assume that the highest risk of pollution would arise if 25 mm of rain were to fall shortly after spraying, and flush 50% of the applied herbicide directly into the water supply. If the total catchment area had been sprayed the resulting level of any of these herbicides in the run-off water, assuming that they had been applied at the recommended rates (2,4–D at 4 kg a.e./ha, 2,4,5–T at 3 kg a.e./ha, asulam at 4 kg a.e./ha) would be less than 10 p.p.m. Thus, to avoid any risk of pollution exceeding 0·01 p.p.m. not more than $\frac{1}{1000}$th of the catchment area should be sprayed in one operation say lasting one week. If all the water from the area to be sprayed is going into a reservoir before consumption, it is safe to assume that considerable dilution will occur, and that up to $\frac{1}{500}$th of the catchment area may be sprayed in one operation.

The catchment area of importance is that area which supplies drainage water to a stream or river at the point at which it is first used as a water supply. This point could be in the forest for local cottages, or it could be many miles downstream. In the former case the catchment area might be quite small—maybe only a few hundred hectares—but usually catchment areas are at least several thousand hectares.

Catchment areas can usually be defined and measured from 1-inch (or 1:50,000) Ordnance Survey maps, but advice can always be obtained from the local water authority. It is unusual for more than $\frac{1}{1000}$th of a catchment area to be sprayed in one week by a ground spraying operation; the main risk of exceeding this suggested limit is from aerial spraying operation.

TABLE 5
PROTECTIVE CLOTHING WHEN USING HERBICIDES IN THE FOREST

CLOTHING	AMMONIUM SULPHAMATE		ATRAZINE AND DALAPON IN WATER		PARAQUAT IN WATER	ASULAM, DICAMBA, 2,4,5-T, 2,4-D OR 2,4-D/2,4,5-T MIXTURE IN WATER		2,4,5-T IN PARAFFIN		2,4,5-T or 2,4-D ULV FORMULATIONS	CHLORTHIAMID OR DICHLOBENIL
	Crystals	Solution at Medium vol.	Medium vol.	Low vol.	Medium vol.	Medium vol.	Low vol.	Medium vol.	Low vol.	Ultra low volume	Granules
BOOTS—WELLINGTON Rubber or Oil-resistant	D	E	E	E	E	E	E	E	E (4)	E (4)	D
TROUSERS OR LEGGINGS Waterproof (1) & (7)	D	E	E	E	E	E	E	E	E (4)	N	D
JACKET Waterproof (7)	N	D	D	E	E	E	E	D/E (4&5)	E (4&5)	N	N
GLOVES Plastic	E	E	E	E	E	E	E	E	E	E	D/E (8)
FACE SHIELD or EYE SHIELD	N	N	N	D	D/E (2)	N	E	D	E	E	N
RESPIRATOR (3)	N	N	N	D	N	N	D	D	D	D	D (8)
HAT or HOOD Waterproof or oilproof	N	N	N	D	N	N	D/E (6)	D (4) &	D/E (4)	E	N
ULTRA LOW VOLUME SUIT (4)	—	—	—	—	—	—	—	—	—	E	—

Key to Table 5

E = Essential. Laid down under the Pesticides Safety Precautions Scheme or considered necessary in relation to working conditions in forestry.

D = Discretionary. Must be available. Such items may reduce discomfort of the operator and so should be freely available but the final choice may rest with him after fair trial.

N = Not considered necessary.

(1.) Waterproof trousers are very uncomfortable to wear due to condensation and heat retention. They should not be worn except when using oil diluent or when there is a high risk of contamination from the spray. Normal contamination from the spray may be more comfortably avoided by wearing waterproof leggings.

(2.) Face shield essential when handling paraquat concentrate but not otherwise.

(3.) A respirator designed to intercept droplets is quite adequate. For sprays in oil, if the smell causes discomfort, a respirator designed to intercept vapour as well as droplets can be used. Special masks are available for ULV spraying.

(4.) The material used for boots and protective clothing needs to be oilproof. Rubber is not oilproof.

(5.) A jacket is essential when using a knapsack sprayer.

(6.) A hat or hood is essential for ground markers during aerial spraying.

(7.) In some circumstances this clothing may also need to be thornproof.

(8.) Paper face mask and gloves may be desired when filling applicators.

CHAPTER 12

NOTES ON MATERIALS AND SOURCES OF SUPPLY

12.1.   The objective in this chapter is to provide enough information to help the user appreciate those characteristics of the herbicides and of the materials used for protective clothing which may be important in practice, and to help them initially to find suppliers. No attempt is made to give a comprehensive description or list suppliers of each herbicide or article of protective clothing.

12.2.   For more detailed descriptions of the herbicides, readers should refer to the Weed Control Handbook, Volume I, 5th Edition—revised (Fryer & Evans, 1970) or the Pesticide Manual, 2nd Edition (Martin, 1971). Fuller descriptions of suitable protective clothing can be found in Forestry Commission Bulletin 48 (Wittering, 1974).

12.3.   Mention of particular suppliers means that their product is considered satisfactory, but does not infer that the products of other suppliers are not equally satisfactory.

HERBICIDES

12.4.   Recommended sources of most herbicides are given in *Approved Products for Farmers and Growers*, published by the Ministry of Agriculture, Fisheries and Food, Publications Branch, Tolcarne Drive, Pinner, Middlesex, HA5 2DT. This list of products approved under the Agricultural Chemicals Approval Scheme is published annually in February. The source of herbicides not included in that publication are given in the appropriate paragraph below.

Ammonium Sulphamate (A.M.S.)

12.5.   A highly soluble, crystalline solid which controls many woody species, and, at high doses, may also be used as a total weedkiller. It is absorbed through leaves and roots, and through freshly exposed surfaces of transportation tissues. It is not, however, well absorbed through bark.

12.6.   It will persist in soils at phytotoxic quantities for periods up to 12 weeks after application, and may damage trees planted too soon after spraying. Ammonium sulphamate has a low mammalian toxicity (L.D.$_{50}$ rats = 3,900 mg/kg).

12.7.   Unfortunately, ammonium sulphamate is highly corrosive to most metals used in sprayers, and although this corrosion can be reduced by using an inhibitor (see para 6.4.3) it is wise only to use it in sprayers in which the spray solution has no contact with metal parts.

Asulam

12.8.   A herbicide originally marketed for control of docks. *Rumex* species in grassland, and which has subsequently been shown to be very active against bracken.

12.9.   The basic compound is not very soluble in water. It is, therefore, formulated as a salt (usually a sodium or potassium salt) and sold as a water soluble concentrate usually containing 40% w/v of asulam. Its mode of entry is mainly through foliage, but it is also taken up from the soil. Its activity against bracken seems to depend on very good translocation from the fronds to the rhizomes.

12.10.   Persistence in the soil is short. However, it often gives long-term control of bracken because of its effect on the regenerative capacity of bracken rhizomes. It is of low mammalian toxicity (L.D.$_{50}$ mice = > 5000 mg/kg).

Atrazine

12.11.   A compound of low solubility in water, which is therefore formulated as a wettable powder (usually containing 50% w/w atrazine) or a granule (4% w/w atrazine). It is absorbed through leaves and from the soil, but mainly the latter. Entry through foliage varies considerably with species, and this may account for the sensitivity of grasses and the tolerance of conifers to this herbicide. At high rates (10–20 kg a.i./ha) atrazine is used for total weed control.

12.12.   Atrazine persists for several months in the soil, and may be detectable even one year after application. However, there are no known residue or leaching problems in forestry. It has a low mammalian toxicity (L.D.$_{50}$ rats = 3080 mg/kg).

Chlorthiamid and Dichlobenil

12.13.   Both herbicides give good control of a wide range of grass and herbaceous broadleaved species, but have little effect on woody species.

12.14.   Chlorthiamid is of rather low solubility in water and dichlobenil is of very low solubility in water. Both herbicides show little activity through foliage, and are formulated as granules containing 7·5% w/w active ingredient. After application, chlorthiamid is broken down in the soil within a few days to dichlobenil. Absorption from the soil may be mainly through underground parts of shoots as the vapour phase of dichlobenil.

12.15.   Chlorthiamid differs from dichlobenil in being very much less volatile and by being rather more soluble in water. It is believed that its lower volatility helps reduce losses of the active ingredient from surface applications and its greater solubility assists incorporation in the soil. Early formulations of dichlobenil were in powder form, and substantial properties of the dichlobenil itself appeared to be lost by volatilisation if dry weather persisted after application, but later formulations in granules (e.g. Casoron G) appear to have overcome this problem.

12.16.   Persistence in the soil is usually between 3 and 6 months. Chlorthiamid is of moderate toxicity (L.D.$_{50}$ rats = 757 mg/kg) and dichlobenil of low toxicity (L.D.$_{50}$ rats = 3160 mg/kg).

Dalapon

12.17.   A very widely used herbicide for controlling grass species. It has little or no effect on broadleaved weeds.

12.18.   The basic compound is rather unstable and it is formulated as a sodium salt, which is a white powder highly soluble in water. This powder is available with or without an added wetter, and contains 74% w/w dalapon. The presence of the wetter usually improves the control of most weeds.

12.19.   Absorption is mainly through foliage, but it is also taken up from the soil. The excellent translocation of dalapon within the plant explains why it is often good at controlling rhizomatous grasses. Dalapon is not tightly held on leaf surfaces, and this may explain why poor results are sometimes associated with rain shortly after application.

12.20. Dalapon does not persist for long in the soil, and toxic levels of the herbicide are rarely present after 3–6 weeks. Mammalian toxicity is very low (L.D.$_{50}$ rats = 7,570 to 9,330 mg/kg).

Glyphosate

12.21. This new herbicide is reputed to give control of a very broad spectrum of weeds, particularly grasses. At the time of writing this publication (mid '74), however, the product had not appeared on the market, although it should soon be available from Monsanto Chemicals Ltd, Monsanto House, 10–18 Victoria Street, London SW1H 0N9.

12.22. Glyphosate is an acid which is slightly soluble in water, but the formulations available are water soluble salts containing about 35% w/v glyphosate.

12.23. Glyphosate appears to be absorbed entirely through aerial portions of plants, especially leaves, and to have no soil activity. It is extremely well translocated; its capacity for movement within plants may enable it to give excellent control of many perennial weeds, even those with substantial underground organs.

12.24. Little information is available on the persistence of glyphosate in the soil. However, it appears to be inactivated in the soil. Its mammalian toxicity is low (L.D.$_{50}$ rats = 4,320 mg/kg).

Paraquat

12.25. Very active on grasses, and will also suppress a wide range of herbaceous broadleaved species.

12.26. Paraquat is available as a water soluble concentrate containing 20% w/v paraquat. It is absorbed mainly by foliage and other photosynthetic parts of plants, but is poorly translocated. Thus kill is usually very local to the site of absorption (contact activity), and this may result in a good top kill of weeds but a rapid recovery. Translocation is better under low temperature and light conditions (e.g. autumn or early spring), and then complete kill of weeds might be obtained.

12.27. Paraquat is persistent in soil, but it is so tightly held by the clay particles in the soil that it is not available to plants. It is of moderate toxicity (L.D.$_{50}$ rats = 155 to 203 mg/kg), although there have been cases in humans of accidental direct intake to lungs or into the blood-stream when it has exhibited greater toxicity.

Propyzamide

12.28. Most grass species are well controlled by propyzamide but many broadleaved weeds are unaffected. Woody species appear to be very tolerant.

12.29. The herbicide has a low solubility in water, and is ineffectively absorbed and translocated by foliage. Uptake seems to be mainly from the soil. Propyzamide is formulated as a wettable powder containing 50% w/v active ingredient, or as a granule containing 4% w/v active ingredient. It is marketed in Britain by Rohm & Haas, (UK) Ltd., Lennig House, 2 Masons Avenue, Croydon, Surrey CR9 3NB.

12.30. The persistence of propyzamide varies enormously with soil temperature. At 5°C or below it is very stable and persists for months, but at 10–15°C or more it is quickly broken down. This probably explains why the most effective application time is late autumn/early winter.

12.31.   There appears to be no residue problem; during the warm summer months soil temperatures are high enough to ensure that the herbicide disappears. Propyzamide is of very low mammalian toxicity (L.D.$_{50}$ rats = 5,620 to 8,350 mg/kg).

2,4–D (2,4–dichlorophenoxyacetic acid) and 2,4,5–T (2,4,5–trichlorophenoxyacetic acid)

12.32.   These two plant-growth regulating type herbicides are widely used throughout the world for controlling herbaceous and woody weeds in many crop situations.

12.33.   Both are acids which are too insoluble to be readily usable and they are made into salts, amines or esters. The salts and amines can be formulated as water-soluble concentrates, which usually contain 50% w/v of the acid. The esters are only adequately soluble in oils and—for application in water—they are formulated with oil and emulsifiers to make a concentrate which will form an emulsion with water. This emulsifiable concentrate usually contains 50% w/v acid. Esters without emulsifiers or other additives are also available ("unformulated" esters). These are suitable for spraying in oil diluents only, in which case the concentrate contains between 75 and 100% w/v of the acid, depending on the type of ester.

12.34.   In practice, 2,4–D ester alone for dilution only in oil is not commercially available. Similarly, 2,4,5–T is not sold as the salt or amine. This is because there are insufficient markets for these formulations, in the United Kingdom.

12.35.   The effectiveness of 2,4–D and 2,4,5–T depends on good translocation within plants, and the initial problem is to get sufficient herbicide into the transportation tissues. With herbaceous or shrubby plants, the only practical way is by foliar applications. Esters are generally more efficiently absorbed through leaf surfaces than salts or amines, although in many non-forest situations, the need for crop tolerance dictates the use of the latter. With trees, stem applications are feasible, but where the method relies on substantial absorption of the herbicide through bark, it is essential to use an oil diluent to transmit the herbicide across this barrier. Thus esters must be used. The only use for salt and amine formulations in forestry appears to be for tree injection techniques; they are thought to be more effective than esters for this method because water soluble forms may be more easily translocated within the plant.

12.36.   Esters are formed by combining the acid with an alcohol. Certain properties of the ester are affected by the type of alcohol used, particularly the volatility of the herbicide. To reduce the risk of excessive volatilisation during and shortly after application, esters having a volatility equal to or lower than the iso-octyl ester are recommended.

12.37.   Both 2,4–D and 2,4,5–T show little activity through the soil. 2,4–D is normally completely broken down within a month, and 2,4,5–T, although rather more persistent, normally breaks down within 6 months. Both compounds are of moderate mammalian toxicity (2,4–D L.D.$_{50}$ rats = 400 to 666 mg/kg; 2,4,5–T L.D.$_{50}$ = 300 to 500 mg/kg).

MARKER DYES, WETTING AGENTS AND OTHER SPRAY ADDITIVES

12.38.   Waxoline Red O.S. to be added to 2,4,5–T in oil for stump spraying can be obtained from I.C.I., Southern Region Sales Office, 81 High Holborn, London WC1V 6NP.

12.39.   "Methyl Violet 5807" to be added to the ammonium sulphamate for stump spraying can be obtained from Hopkins & Williams, Freshwater Road, Chadwell Heath, Essex and "Hexacol Violet B.N.P. Extra" from L. J. Pointing & Sons Ltd., Dyestuffs Manufacturers, Hexham, Northumberland.

12.40.   Several non-ionic wetting agents are suitable for adding to dalapon or ammonium sulphamate solution. These include "Nonidet" made by Shellstar Ltd., and "Agral 90" made by Plant Protection Ltd.

12.41.   Sodium benzoate (corrosion inhibitor for use with ammonium sulphamate in brass spraying equipment) can be obtained from many local chemists.

PROTECTIVE CLOTHING

Wellington Boots (oil-proof)

12.42.   These should be knee length and lined. Unlined boots are too easily pierced by sharp objects in the forest. Found to be satisfactory are "Oil-Boy" (article 65769 08) from Semperit Rubber Manufacturing Co. Ltd., Wexham Road, Slough, Bucks.

Jacket and Trousers

12.43.   These need to be tough and thorn-proof to avoid tearing in the forest. They need to be oil-resistant for oil-borne sprays and water-proof for other sprays. A compromise is necessary between types that give complete protection but give insufficient ventilation and types giving good ventilation but only marginally acceptable protection.

12.44.   For oil-borne sprays "Jalites All-purpose Polyurethane Suit" (green) from Abridge Overalls Ltd., Burgess Hill, Sussex, is considered satisfactory. For water-borne sprays the "Polymac Suit" (green) from Watsons (Newburgh) Ltd., Newburgh, Fife, Scotland, gives adequate protection.

Protective Clothing for Ultra Low Volume Spraying

12.45.   Because there is negligible risk of gross contamination of the operator's protective clothing in ULV spraying and it is only necessary to contend with very small, light droplets, it has been possible to develop a special protective "suit" for this operation. The special feature of this suit is that it allows some movement of air through the material, but still prevents liquids (as small droplets) from passing. This suit is called the Jalite ULV suit from Abridge Overalls Ltd., Burgess Hill, Sussex.

12.46.   Worn with oil-proof Wellington boots and gloves, this suit protects everything except the face. Because of the small droplets produced by ULV sprayers protection of the nasal passages is considered essential, and for this the "3M Brand Non-toxic Particle Mask No. 8500" from Herts Packaging Co. Ltd., 53 London Road, St. Albans, Herts, are suitable.

Gloves and Barrier Cream

12.47.   Unlined gloves are theoretically better because if lining is contaminated with herbicide, the risk of skin damage is increased by continual contact with the contaminated lining. However, unlined gloves are usually too weak, and so a compromise of using unlined gloves just for handling the concentrate, and lined gloves plus barrier cream for handling spray solutions is recommended.

12.48.   For concentrates, unlined disposable gloves are suitable, like the polythene gloves sold by The Polythene Glove Co., Rydal House, Copse Road, Fleetwood, Lancs FY7 6RP. For spray solutions, the "No. 540 Plastochrome Glove" from James North & Son, Ltd., P.O. Box 3, Hyde, Cheshire, plus "Rosalex No. 10" barrier cream from Rosalex Ltd., Road One, Industrial Estate, Winsford, Cheshire, are suitable.

12.49.   This barrier cream can also be used to protect other exposed skin areas.

Goggles and Face Shields

12.50. Not all goggles will fit the face properly when a respirator is being worn. It is better to use goggles which can be worn with a respirator to cover those occasions when the spraying operator requires both.

12.51. Goggles in particular, and to some extent face shields, are prone to "misting" during work. Therefore, some ventilation of the inside is necessary to prevent this.

12.52. The "Clearways Face Shield" from Safety Products Ltd., Holmthorpe Avenue, Redhill, Surrey, and Face Shield FS/1318/BW from James North Ltd., Hyde, Cheshire, have been found to be satisfactory. The "Panoramette Chemical Goggle" from Pyrene Panorama Ltd, Hanworth Air Park, Feltham, Middlesex, is a satisfactory goggle, but cannot be worn with the suggested respirators (below). The best Eye protection found so far for wearing with these respirators is the "Eye Shield type 6638" from Industrial Glove Co., Ltd., Nailsea, Somerset, BS18 2BX.

Respirators

12.53. These are regarded as essential for all Ultra Low Volume (ULV) spraying. For other types of work, none is considered essential, but operators may request a respirator for comfort. Some herbicides are capable of causing discomfort because they or the diluent used are volatile (e.g. 2,4–D and 2,4,5–T, especially in paraffin diluent). For these, respirators with filters that extract vapours from the air are necessary. For other herbicides with negligible volatility (e.g. paraquat), respirators fitted with particulate filters are all that is required.

12.54. For protection against vapours (and fine droplets) the "Toxigard Agricultural Respirator Type QR 3086" fitted with two "RC 86 Cartridges" from Mattay & Co. Ltd., Mattay House, 2 Higher Road, Liverpool L25 OQQ is satisfactory. For fine droplets only, the "Baxter Pneu-Seal Dust Mask" with dust cartridge BS 2091 from The Leyland & Birmingham Rubber Co. Ltd., Leyland, Preston, Lancs., PR5 1UP is satisfactory. For ULV spraying, where droplet size is fairly uniform, the 3M Brand Non-toxic Particle Mask No. 8500 from Herts Packaging Co. Ltd., is considered satisfactory (see para 12.46).

Head Wear

12.55. Only essential for aerial spraying (ground markers), but can be requested for mistblowing (for ULV see separate head below). An oil-proof Sou'wester, such as the types sold by James North & Son Ltd., and Abridge Overall Ltd., are satisfactory (for addresses see para 12.48 and 12.44 above).

Cleaning Facilities

12.56. Soap and water are the best, but not always practical in the forest. A hand cleansing cream plus paper towels are recommended as a suitable alternative.

12.57. "Waterless Skin Cleaner No. 44" from Izal Ltd., Thorncliffe, Chapeltown, Sheffield, S30 4YP, plus paper towels from Jeyes UK Ltd., Brunel Way, Thetford, Norfolk, are suitable.

ACKNOWLEDGEMENTS

Without the forerunner of this booklet, Forestry Commission Leaflet No. 51 by Mr. J. R. Aldhous (Second Edition, 1969, now out of print), it would have been immeasurably more difficult to write. My thanks are also due to Mr. Aldhous for his direct help in preparing the manuscript.

I would also like to thank all my other Forestry Commission colleagues who helped with the manuscript, and particularly Mrs. A. Walters who typed and retyped the typescript as necessary.

The photograph for the front cover (right) was kindly provided by British Petroleum Ltd. The photograph for plate 8 was provided by Mr. W. O. Wittering of the Forestry Commission Work Study Branch. All other photographs were provided from the Forestry Commission photographic library at Alice Holt.

REFERENCES

ALDHOUS, J. R. (1965). Weed Control in the Forest. *Rep. Forest Res., Lond.* for 1964. p. 35.

ALDHOUS, J. R. (1967). 2,4-D residues in water following spraying in a Scottish forest. *Weed Research* 7(3), 239–241.

ALDHOUS, J. R. (1972). *Nursery Practice*. Bulletin 43, For. Commn. Lond. HMSO £1·50.

AGRICULTURE, MINISTRY OF. *List of Approved Products*. Agricultural Chemicals Approvals Scheme, M.A.F.F., Middlesex, Britain.

ANON (1969). *Recommended Common Names for Pesticides*. British Standard 1831, British Standards Institute, London.

ANON (1969). Forest Weed Control (Side effects of Chemicals). *Rep. Forest Res. Lond.* for 1969, 80–83.

AUDUS, L. J. (1960). Microbiological breakdown of herbicides in the soils. *Herbicides in the Soil*. Blackwell, Oxford.

BURSCHEL, P. (1963). Das Verhalten der forstlich wichtigen Herbiziden im Boden. *Forstarchiv* 34(9), 221–233.

BROWN, R. M. and MACKENZIE, J. M. (1971). Forest Weed Control (Woody Weed Control). *Rep. Forest Res., Lond.* for 1971, p. 55.

BROWN, R. M., ROGERS, E. V. and THOMSON, J. H. (1973). Ultra Low Volume (ULV) Spraying. *Forestry and Home Grown Timber* Vol. II, No. VI.

CONNELL, C. A. and COUSINS, D. A. (1969). Practical Developments in the Use of Chemicals for Forest Fire Control. *Forestry* 42(2) 1969, 119–132.

EVERARD, J. E. *Fertilisers in the establishment of conifers in Wales and Southern Britain*. Booklet 41. For. Commn., Lond. (HMSO £1·25).

ERSKINE, D. S. C. (1968). Experimental work in the control of bracken, 1958 to 1968. *Proc. 9th Brit. Weed Cont. Conf.*, 1968. 488–492.

FRYER, J. D. and EVANS, S. A. (1970). *Weed Control Handbook, Volume I—Principles*. 5th Edition (Revised), Blackwell, Oxford and Edinburgh.

FRYER, J. D. and MAKEPEACE, R. J. (1972). *Weed Control Handbook, Volume II—Recommendations*. 7th Edition, Blackwell, Oxford and Edinburgh.

HANDLEY, W. R. C. (1963). *Mycorrhizal Associations and Calluna Heathland Afforestation*, Bull. 36 For. Commn., (HMSO) Lond.

HOLMES, G. D. (1957). Control of Woody Species. *Rep. Forest Res., Lond.* for 1956, p. 47.

HOLMES, G. D. (1963). Weed Control in the Forest. *Rep. Forest Res., Lond.* for 1962, p. 37/38.

HOLMES, G. D. and FOURT, D. F. (1961). Chemical Aids to Forest Fire Control. *Forestry* 36(1) 1963, 91–108.

LINDIN, G. MULLER, A. and SCHICKE, P. (1963). Versuche zur Frage der möglichen Grundwassergefährdung mit 2,4,5-T in Dieselöl. *Zeitung der Pflanzenkrankheit* 70(7), 399–407.

MARTIN, D. J. (1968). Bracken control trials in West Scotland. *Proc. 9th Brit. Weed Control Conf.*, 1968. 1242–1244.

MARTIN, H. (1971). *Pesticide Manual*. 2nd Edition, British Crop Protection Council.

TURNER, D. J. (*in draft*). *The Safety of 2,4,5-T and 2,4-D in British Forestry*. (Forest Record For. Commn. Lond.).

WALLIS, K. E. *Chemical Control of Heather in Plantations*. Leaflet 64. Forestry Commission. HMSO, London (*in press*).

WITTERING, W. O. *Weeding in the Forest*. Bull. 48, For. Commn., Lond. (HMSO £2·10).

APPENDIX A

PUBLICATIONS REPORTING FORESTRY COMMISSION RESEARCH

ALDHOUS, J. R. (1964). Control of bracken (*Pteridium aquilinum* L.) with dicamba. *Proc. 7th Brit. Weed Cont. Conf.*, 896–898.

ALDHOUS, J. R. (1964). The effect of paraquat, 2,6–dichlorothiobenzamide and 4–amino–3,5,6–trichloro-picolinic acid ("Tordon") on species planted in the forest. *Proc. 7th Brit. Weed Cont. Conf.*, 267–275.

ALDHOUS, J. R. (1966). Bracken control in forestry with dicamba, picloram and chlorthiamid. *Proc. 8th Brit. Weed Cont. Conf.*, 150–159.

ALDHOUS, J. R. (1966). Control of *Rhododendron ponticum* in forest plantations. *Proc. 8th Brit. Weed Cont. Conf.*, 160–166.

ALDHOUS, J. R. (1967). 2,4–D residues in water following spraying in a Scottish forest. *Weed Research*, 7(3), 239–241.

ALDHOUS, J. R. (1967). *Progress Report on Chlorthiamid ("Prefix") in Forestry*, 1962–1966. For. Commn. Research and Development Paper No. 49.

ALDHOUS, J. R. (1967). *Review of Practice and Research in Weed Control in Forestry in Great Britain.* For. Commn. Research and Development Paper No. 40.

ALDHOUS, J. R. (1969). Aircraft in British Forestry. *Quart. J. For.* **63**(2) 105–113.

ALDHOUS, J. R. (1969). Aerial Spraying of Forests—Notes for Landowners. *Quart. J. For.* **63**(2), 152–155.

ALDHOUS, J. R. (1970). UK: trends in forest weed control. *Span* **13**(1), 21–23.

BROWN, R. M. (1968). The effect of chlorthiamid on young forest trees. *Proc. 9th Brit. Weed Cont. Conf.*, 975–980.

BROWN, R. M. (1968). Further experiments on the control of bracken in forestry with dicamba. *Proc. 9th Brit. Weed Cont. Conf.*, 981–988.

BROWN, R. M. (1969). Herbicides in Forestry. *J. Sci. Fd. Agric.* **20**, 21–24.

BROWN, R. M. (1970). Atrazine and ametryne for grass weed control in British Forestry. *Proc. 10th Brit. Weed Cont. Conf.*, 718–726.

BROWN, R. M. (1972). Further trials with atrazine for controlling grass weeds in British Forestry. *Proc. 11th Brit. Weed Cont. Conf.*, 591–600.

BROWN, R. M. and Thomson, J. H. (1974). Trials of ULV Applications of Herbicides in British Forestry. *Br. Crop Prot. Counc. Monogr. No. 11,*   .

FORESTRY COMMISSION (1952 to 1974). Annual Research Reports for years ending March 1952 to March 1974 (one report per year). *Rep. Forest Res., Lond.* for — (year).

HOLMES, G. D. and BARNSLEY, G. E. (1953). The Chemical Control of *Calluna vulgaris. Proc. 1st Brit. Weed Cont. Conf.* 289–296.

HOLMES, G. D. (1956). Experiments in the Chemical control of *Rhododendron ponticum. Proc. 3rd Brit. Weed Cont. Conf.* 723–730.

ROGERS, E. V. (1974). The Selection and Development of Equipment and Methods for ULV Herbicide Spraying in Forestry. *Br. Crop Prot. Counc. Monogr. No. 11,*   .

STOAKLEY, J. T. (1962). Chemical control of coppice and scrub canopy at Bernwood Forest and Wheddon Chase. *Quart. J. For.* **56**(4) 276–292.

WALLIS, K. E. *Chemical Control of Heather.* Forestry Commission Leaflet 64. (H.M.S.O., *in press*).

APPENDIX B

GLOSSARY OF TECHNICAL TERMS

Absorption:	Uptake of a herbicide into a plant.
Acid equivalent (a.e.):	Amount of active ingredient expressed in terms of the parent acid.
Active ingredient (a.i.):	Constituent of a commercial formulation which is the compound with phytotoxic properties.

Application (Method):
Band:	When the herbicide is applied as a band, normally straddling the crop row.
Directed:	When the herbicide is directed towards the ground or weeds to minimise contact with the crop.
Guarded:	When the crop plants are protected from direct contact with the spray by a guard or guards, usually attached to the sprayer.
Overall:	When the spray is applied uniformly over the whole area, as opposed to band or spot application.
Overhead:	When the spray is applied over the crop, as opposed to directed or guarded application.
Spot:	When the herbicide is applied to individual, small patches of weed, usually immediately around the crop trees.
Stem:	Where the herbicide is applied to the stem or trunk of a tree. For the different methods of doing this see paras 3.19 to 3.28.

Application (Time):
Pre-planting:	Before the crop is planted.
Post-planting:	After the crop is planted.
Carrier:	Liquid or solid material added as a diluent to a chemical to facilitate its distribution.
Diluent:	*see* Carrier.
Dose:	Amount (weight or volume) of active herbicide per unit area, or per plant etc. The term *rate* is frequently used in error.
Emulsifier:	Substance which can be used to facilitate the formation of an emulsion of one liquid with another.
Emulsion:	Mixture in which very small droplets of one liquid are suspended in another liquid (e.g. oil droplets in water).

Formulation:
	(1) Process by which herbicide compounds are prepared for practical use.
	(2) Preparation containing a herbicide in a form suitable for practical use.
Emulsifiable concentrate:	Concentrated solution of a herbicide and an emulsifier is an organic solvent which will form an emulsion spontaneously when added to water with agitation.
Granular:	Type of formulation for dry application consisting of granules which serve as a carrier for the herbicide.
Wettable powder:	Type of formulation for spray application in which a herbicide is mixed with an inert carrier, the product being finely ground, and with a surface-acting agent added so that it will form a suspension when agitated with water.

Hectare sprayed or treated:	Treated area, excluding all unsprayed or untreated patches or strips, which adds up to one hectare. Most strip or spot applications cover only 20 to 50% of the gross area, but for recommendation purposes it is necessary to refer to the dose of herbicide required to cover a whole hectare.
Herbicide:	Chemical which can kill or suppress the growth of certain plants.
Contact herbicide:	Chemical that only kills plant tissues very locally to its point of contact with the plant.
Residual herbicide:	Chemical applied to the soil where it remains active for at least several weeks.
Selective herbicide:	Chemical which, if used appropriately, will control some plant species but will not affect others.
Translocated herbicide:	Chemical which, after uptake, is moved within the plant and can affect parts of the plant remote from the point of application.
L.D. 50:	The single dose (lethal dose), in milligrammes per kilogramme of body weight (mg/kg), which will kill 50% of the animals under test.
Pesticide:	Term covering herbicides, insecticides, fungicides, nemacides etc.
Stickers:	Substance added to a spray solution to help retain it on leaf surface.
Surfactants or Surface-active agents:	Substances which, when added to the spray solution, alter the surface tension of the spray liquid and increase its ability to wet the surface of leaves.
Volume of application:	Amount of spray solution (diluent plus herbicide) applied per unit area. The following definitions are used.
High volume:	Volume of application over 700 litres/ha.
Medium volume:	Volume of application between 200 to 700 litres/ha.
Low volume:	Volume of application between 90 to 199 litres/ha.
Very low volume:	Volume of application between 20 and 89 litres/ha.
Ultra low volume:	Volume of application less than 20 litres/ha.
Weight per weight (w/w):	Means of expressing the amount of active ingredient in a commercial formulation by relating this amount, by weight, to the weight of the formulation (e.g. 20% w/w = 0·2 kg in every 1·0 kg of formulation).
Weight per volume (w/v):	Means of expressing the amount of active ingredient in a commercial formulation by relating this amount, by weight, to the volume of the formulation, *assuming that 1·0 litre of every formulation weighs 1·0 kg* (e.g. 20% w/v = 0·2 kg in every litre of formulation).
Wetter or wetting agent:	*see* surfactant.

Printed in Scotland by Her Majesty's Stationery Office at HMSO Press, Edinburgh
Dd 289206 K40 8/75 (12487)